How We Became Our Data

How We Became Our Data

A Genealogy of the Informational Person

COLIN KOOPMAN

The University of Chicago Press
Chicago and London

The University of Chicago Press, Chicago 60637
The University of Chicago Press, Ltd., London
© 2019 Colin Koopman
All rights reserved. No part of this book may be used or reproduced in any manner
whatsoever without written permission, except in the case of brief quotations in
critical articles and reviews. For more information, contact the University of Chicago
Press, 1427 E. 60th St., Chicago, IL 60637.
Published 2019
Printed in the United States of America

28 27 26 25 24 23 22 21 20 19   1 2 3 4 5

ISBN-13: 978-0-226-62644-4 (cloth)
ISBN-13: 978-0-226-62658-1 (paper)
ISBN-13: 978-0-226-62661-1 (e-book)
DOI: https://doi.org/10.7208/chicago/9780226626611.001.0001

Library of Congress Cataloging-in-Publication Data

Names: Koopman, Colin, author.
Title: How we became our data : a genealogy of the informational person /
    Colin Koopman.
Description: Chicago : The University of Chicago Press, 2019. |
    Includes bibliographical references and index.
Identifiers: LCCN 2018048197 | ISBN 9780226626444 (cloth : alk. paper) |
    ISBN 9780226626581 (pbk. : alk. paper) | ISBN 9780226626611 (e-book)
Subjects: LCSH: Information science—Social aspects—United States. |
    Information society—United States—Psychological aspects. | Information
    technology—Social aspects—United States.
Classification: LCC Z665 .K787 2019 | DDC 303.48/33—dc23
LC record available at https://lccn.loc.gov/2018048197

♾ This paper meets the requirements of ANSI/NISO Z39.48-1992 (Permanence of Paper).

Contents

Preface

We are swaddled in data. From cradle to grave, we accumulate an abundance of information. The list of the databases we populate is long, growing, and familiar: search engine and web browsing histories, social media registries, marketing and advertising profiles, predictive policing analyses, and suspected terrorist lists. Add to these the still-multiplying legacy systems of health records, education transcripts, financial data, insurance profiles, and our government records at the federal, state, and county levels, all bookended by our birth and death certificates.

We are not as separable from these data as we like to think. Our data do not simply point at who we already were before information systems were constructed. Rather, our information composes significant parts of our very selves. Data are active participants in our making. The formats structuring data help shape who we are.

Like any swaddling, our data are both constraining and comforting. Information opens up possibilities for what we can be. Yet it also forecloses possibilities for what we cannot be. Like a newborn whose movement is restrained by tightly tucked fabric, we are soothed by data that calm us into stillness and eventually into unthinking sleep. We learn to live within the formats of our information and eventually come to depend on them. These dependencies are dangerous, because where we all find ourselves formatted, it is possible for us to be formatted unequally, even unjustly, and without anyone, including those who most benefit, ever intending it.

Does our being stored away in databases really matter? Does our being defined by data actually make a difference to how we live? Is there really a politics and ethics of data itself? Even the clearest instances of data politics are often met with skepticism. Consider our most unsettling data scandals of late: Edward Snowden's whistleblowing on massive state surveillance at the National Security Agency, consumer credit rating firm Equifax's exposure of the personal information of almost half of Americans, and the tailored delivery of fake news to millions of Facebook users during a presidential election.

The response to Snowden's exposé is particularly noteworthy. The shock we immediately felt when his story broke quickly dissipated into exactly that indifference that we could not have possibly felt as our first reaction. Almost nobody who heard about these revelations reacted with casual unconcern, and yet almost everybody ended up barely caring, or not knowing how to care. Is widespread indifference perhaps the deeper scandal of our new scale and style of surveillance?

We need to ask why so many of us keep retreating into apathy, and even outright cynicism, about the politics and ethics of information. We need to understand why we do not question, and why we even eagerly participate in, projects of government data harvesting, corporate data collection, and a raft of programs designed to store and analyze every flake of data dandruff we cannot help but leave behind in nearly everything we do.

One reason we want to be indifferent to the politics and ethics of data concerns a deep-seated conception we have of ourselves. Thinking of ourselves as our data seems abstract, disconnected, and remote. Our lives do not feel like information. And so we console ourselves with solemn insistences like "I am not a number." Yet the next day, or even the next hour, we unblinkingly present ourselves to the clerk or the computer as just that: a number—for instance, a bank account number, an employee number, a student number, or any other number of numbers that we all in actual fact wield on a regular basis. We want to believe that who we are is one thing, and that all our many numbers are something else. "I am here," we proudly state, fingers thumping our chests or pointing at our skulls. And our data is over there, we maintain, gesturing to a vague elsewhere. But what are we gesturing toward if we know that those numbers always point straight back to us?

Our skeptical reactions are understandable. Many of the massive reams of personalizing data we confront today did not exist a decade ago. It is easy for us to insist that we are not our social media profiles since we so clearly lived for so long without them. Understandable as this reaction is, however, it is perilously misguided. This book proposes instead that we take quite seriously the problems posed to us by the politics of information. There are two steps to my proposal.

The first step involves considering the consequences of the plain fact that we find ourselves enrolled in a thousand databases. Who are we without all these identifiers, numbers, and other bits stored away in countless many data warehouses? Who could you be without your data points? What could you do? Make a list of everything you have done this week. What on that list required you to offer up your data in doing it? If you could not have done those things this week (transact with the bank, check your email, see your grades or those of your students, use your keycard to access your building), what other things would you not be able to do next week? How long before you would lose your job? Miss a few rent or mortgage payments? Suffer from lack of medical care? How many weeks before you would be wholly reliant on others for maintaining even your most basic life functions as eating and sheltering? We like to believe that we are not our data. But we inhabit lives that rely on data in nearly every act we perform. We are therefore our data as much as we are anything else. We are many things, of course. But we are our data too.

If the first step of my proposal is to come to terms with the fact that we live and act through our data, then the second step is to ask how we have become our data. How were our current dependencies on data formed? What from our past persists in conditioning our present lives of data?

Skepticism here redoubles. Information would have us believe that it has no history. Information technology is purveyed as always new, and as desirable precisely because it is new. Avant-gardism about data is frequently assumed by both its biggest boosters (in the tech sector and the intelligence state) and its most scrupulous critics (among data activists and my fellow academics). We need to disrupt these ahistorical assumptions about information.

Accepting information as ahistorical facilitates our tendency to take information technologies as closed, locked, and unchangeable. In using information systems of all kinds, from tiny devices to enterprise platforms, we familiarly find ourselves feeling that we must yield to the technology. We fill out the form the way it wants to be completed. We produce the numbers that the system solicits. We do exactly what the computer instructs us to— and most especially when we cannot get it to do what we want. And yet the designs of data are anything but immutable forms that we must acquiesce to. In investigating the history of information, we establish contact with the mobility and manipulability of data technology. For we can find in that history a set of moments when data was not yet closed, but rather glaringly open to contestation and recomposition.

The pasts of data show us that information technology is ever open to revision. Yet those same pasts also show that the weight which information

technology bears is not newly gained. The history of information's importance is longer than is customarily conveyed in our stories about the last three
decades of social media, internet surfing, and personal computing. It is also
longer than that other story we often hear about the birth of an era of information back in the mid-twentieth century. We started becoming informational persons in a turbulent moment prior to our entry into "the information
society" of "the information age" in which we remain. Transformable data is
deeply entrenched and as a result not easily transformed.

The genealogy explored in this book locates the emergence of informational personhood in an unsettled period running from the mid-1910s to
the mid-1930s—this was the long 1920s, inclusive of its dawning and setting.
Amid the riotous moment of a roaring decade, we were formatted into data
of uncountably many kinds: birth certificates, psychological assessments,
education records, financial profiles, the production of a sizable racialized
data apparatus, and so much other informational accoutrement that we have
for so long now simply taken for granted. These century-old formats remain
with us today. They persist in the latest information technologies that, new as
they are, depend on older techniques for their deepest infrastructure.

Older information infrastructures matter much for data today. This is not
a book that directly examines our most recent contemporary technologies.
This is not a book about our technological tomorrows or any other near future toward which our present may be hurtling. Rather, this is a book about
the turbulent pasts out of which were formed our present enthusiasms for
bigger-than-ever data and better-than-ever informatics. This is a book about
the persistent histories in which so much of what is promised for tomorrow
is already embedded.

If we have recently been reformatted once again by newer high-tech platforms heralding newer highest heights, these latest formats cut deeper into a
groove that was first carved a century ago. Both moments matter—and both
matter much. For what was deployed just last year already depends on what
was designed one hundred years ago. That is why the history of how we became our data matters for how we are our data today.

Informational Persons and Our Information Politics

A Life of Data

In the late years of his life, Viennese polymath Otto Neurath worked on a project of self-documentation he described as "a visual autobiography." His virtuosic effort in self-description remained unfinished at his death in 1945, and the drafts he left behind were not collated and published in full until 2010. Neurath envisioned a book composed of interwoven text and image to communicate the multimedia adventure of his life. The culmination of the autobiography was to be a chapter devoted to the genesis of Isotype, a universal pictorial language Neurath had been developing in multiple collaborative venues since 1923. The draft autobiography described Isotype as "a common visual basis of information."[1]

Among the rich visual reproductions of Isotype charts in the autobiography's draft is Neurath's remembrance of a department store exhibition he prepared on Rembrandt. The conceit of the exhibition, just one of many such Isotype projects, was to convey to the public the life of a great artist through a universal, or intertranslatable, symbolic iconography—a life told, in short, in information. Neurath's presentation of the Rembrandt exhibition in the course of his own autobiography suggests the provocative possibility of a life composed in Isotype.[2]

Neurath envisioned Isotype as an "international picture language which would be able to bring together all kinds of people."[3] It was to be a vehicle for communicative unification. Isotype was one of many grand unifying ambitions to which Neurath dedicated his life. In multiple domains, from communication and information design to economics and social planning to the natural and social sciences, Neurath sought unity and universality. Unification even inflected Neurath's choice of presentation of Isotype, as evidenced in a hand-drawn diagram through which he sought to convey the "genealogy" of

Isotype on the basis of a multiplicitous convergence of sciences. The apogee of Neurath's vision of unification was perhaps the unity of science project for which he is best known among contemporary scholars of philosophy.[4]

Neurath's thought expressed a specific yearning for a form of universality that can be retroactively regarded as encapsulating the universality of information itself. This too is clarified in Neurath's work on Isotype.[5] The final sentence of his 1936 book *International Picture Language* offers a bold statement to this effect: "What the science of reasoning has done to make possible such a uniting of the sciences and to give one word language to all the special sciences, the ISOTYPE system has done to make possible one language of pictures which will give the same sort of help to the eye for all the special sciences and for persons of all nations."[6] He would articulate his project's demands in similar terms in a 1937 article: "We need a new way to convey information, a method which is simple to teach and learn, and at the same time comprehensive and exact. One solution is Isotype."[7] This latter statement is certainly more measured, but it is equally remarkable for how it explicitly couches the issue as a matter of information.

Such yearnings for informational universality persist today with arguably greater vigor and demonstrably greater success. One poignant example vis-à-vis Neurath's language unification ambition is Google Translate, presented as a project that "connects people in communities around the world."[8] Translate is, moreover, just one of a suite of Google efforts that help the company realize its stated mission to "organize the world's information and make it universally accessible and useful."[9] That this and so many other yearnings for universal informatics are preeminent today prompts important questions whose answers remain elusive.

What made it possible for Neurath, and so many others of his generation, to conceive of lives and histories in terms of intertranslatable symbols—that is, as information? What made it possible for proponents of cybernetics in the same decade to seriously propose that the universe itself is essentially informational in its nature? Why were informational ambitions so compelling in those heady midcentury years such that there could soon emerge in their wake the "information society" and "information age" within which we still live today? What were the earlier historical conditions of these midcentury moments such that these very moments could later become the historical conditions of what is now our present?

From the historical perspective that bends an arc from the present back to the past, what is perhaps most remarkable about Neurath's dream of a life composed of data is that it soon became the working reality of cadres of data technicians of all stripes, from computer software engineers to corporate

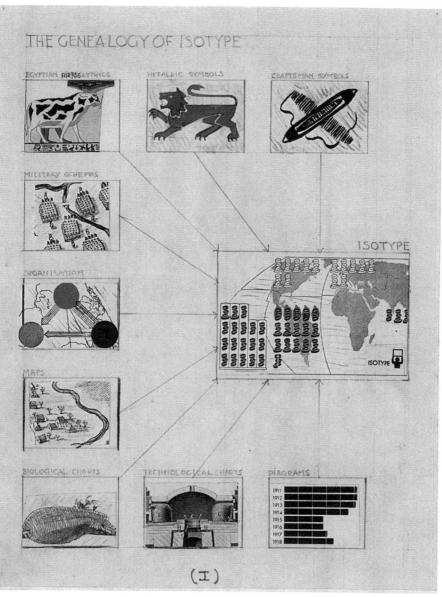

FIG. 0.1 Otto Neurath's hand-drawn "Genealogy of Isotype," ca. 1946–47, from the Otto and Marie Neurath Isotype Collection at the University of Reading (IC 3.2/86)

marketing analysts to government intelligence specialists to garage wizards inventing tomorrow's technological infrastructure. It then became in turn the everyday actuality of us all. From one information system to the next, and across each, we are inscribed, processed, and reproduced as subjects of data, or what I call *informational persons*. The extent to which we informational persons are so widely formatted into our data suggests the high stakes of our datafication and its concomitant politics of information, or what I call our *infopolitics*.

Informational Persons

We are subject to vast amounts of personal data that others attach to us and that we in turn regularly reattach to ourselves. These data points have become important to who we are. More precisely, they are important to who we have become. To understand their importance, consider the following hypothetical scenario of a kind of nightmare that could only come to pass in a milieu such as ours.

What would it be like to permanently lose access to all your data all at once? Beyond just simple informational identity theft or misplaced data records, envision a scenario of permanent personal data deletion. The prospect, when thought through, is truly frightening. What would you do if you somehow became permanently detached from all your personal data? What could you do if you somehow became permanently unrepresentable by all data systems?

This is precisely what the informational person dreads most: the permanent and irreversible erasure of the entirety of their personal information and therefore their entire informational identity. No driver's license, no passport, no bank account number, no credit report, no college transcripts, no employment contract, no medical insurance card, no health records, and, at the bottom of them all, no registered certificate of birth. The scenario is chilling: everyone around you well attached to their data while you are dataless, informationless, and as a result truly helpless. What would you make of yourself? What could others make of you? What would the bureaucracy be able to do when you petition it with your plight, given the fact that no bureaucracy can address a subject as other than their information? The bureaucrats, like your family members, could surely recall your visage and your voice. But they would have no way of addressing you from one day to the next, of recording you in their databases, of numbering or naming you, and so no way at all to deal with you on anything approaching a consistent basis. You could not even receive special support through special court orders because, completely unrepresentable as information, you would have no way of being registered into a court, for that would require rendering you into the data from which you have been detached.

Some parts of you would at first remain well intact: your body, your mind, much of your memory, and your talents (though those requiring certification of any kind would be useless to you). But of what lasting value could these be in a world like ours? What could you make of yourself with them? What could others begin to make of you on the basis of your mind and your body deprived of your data?

This nightmarish scenario helps us recognize that the loss of one's informational selfhood would entirely debilitate our sense of self today, even if it would leave intact other aspects of who we are. Our information is today so deeply woven into who we are that were we to be deprived of it, we could no longer be the persons we once so effortlessly were. If this scenario is chilling, then, it is a chill we feel only within a particular historical context. Were you to propose this scenario to a farmer tending his fields some three hundred years ago, the entire thought experiment would be enormously difficult for him to even envision; that farmer would have tilled and harvested in a time when almost none of his neighbors possessed any of the informational accoutrement we rightly fear being deprived of. Those few neighbors who did make use of records of their selves would in almost every instance be wealthy or titled, and almost certainly owners of valuable property. But the average person would not be so much chilled at the prospect as rather simply puzzled at what is even being proposed.

INFORMATION'S PERSONS

Cornelia Vismann observes that in the nineteenth century people began to "produce themselves by administering themselves, by establishing a feedback with their own actions. . . . The demand for accountability applies not only to merchants and their shenanigans but to the most banal chores and most secret ideas of an individual, the bookkeeping practices common to business offices are transformed into diaries, autobiographies, and other such accounts."[10] Vismann's account emphasizes the confessional technologies made famous in Michel Foucault's genealogies of the modern sexual subject.[11] Complementing this were a whole range of less literary and more mundane materials for self-formation. Ian Hacking, also writing of the nineteenth century, describes practices and institutions that "brought a new kind of man into being, the man whose essence was plotted by a thousand numbers."[12] If Vismann emphasizes the confessional vectors of Foucault's genealogy of sexuality, then

Hacking emphasizes its statistical vectors.[13] The story told by all three of these philosopher-historians concerns the formation in the nineteenth century of what Foucault's work describes as biopolitical and disciplinary subjectivity.

Not only are such subjects still with us today as crucial aspects of contemporary subjectivity, but they also set the stage for other new kinds of persons that soon began to be born. If the confessing and statistical subjects were key figures of the nineteenth century, then the early twentieth century gave rise to then-new subjects of information that continue to be presented to us today as still new. A decisive weight was tipped in those first decades of the twentieth century. Information began to precede the person. It became possible for information to draw up persons as if out of nowhere. We became coddled from cradle to coffin by so many check boxes on so many scraps of paper. We began being born onto forms: the ubiquitous birth certificate certifying the inauguration of a lifelong paper trail that would outlive even the eventualities of our death certificates.

The contemporary aftermath of these checkboxes and forms in their digitized variations—and scaled up to increasingly "big" levels of computation—is eminently familiar to us today. It has been the focus of much recent scholarship, including by scholars who, like me, seek to redeploy Foucault's histories of subjectivity for the purposes of conceptualizing contemporary informational personhood. Natasha Dow Schüll's analyses of contemporary self-tracking technologies draw attention to the granular and intensive "datafication" at work in the "tracking and coding of bodies of acts," all giving rise to "a database self whose truth lies in scattered points, associations, and dynamic accretions."[14] Bernard Harcourt argues that today "data points constitute a new virtual identity, a digital self that is now more tangible, authoritative, and demonstrable, more fixed and provable than our analog selves."[15]

Vismann's files, Hacking's statistics, and the newer data constellations of Schüll and Harcourt all exemplify Lisa Gitelman's crucial claim that "new inscriptions signal new subjectivities."[16] To these and other genealogies this book adds an account of an unexplored episode in the history of our contemporary subjectivities.[17] This account is drawn out of a historical period that too frequently falls through the cracks in our consideration of information technologies: a period in the early twentieth century in which was born the informational person whose historical timing is situated between twenty-first century digital profiles and the statisticalized populations and confessing individuals of the nineteenth century.

But before excavating a history of our present from this neglected period of a century ago, I need to scrutinize that present itself. Who is the informational person that our information technologies have coaxed us to become?

Consider a contemporary emblem of the informational person that in only a few short years has become utterly banal: the social media profile. These profiles are already easily navigated by users even as they are regularly repurposed by developers eager to offer new applications for presenting ourselves. The genre pivots around a remarkably simple idea: a flat two-dimensional space on which an array of data are composed as a presentation of persons. Each element in the informational composition ostensibly speaks to some aspect of our selfhood, and the overall ensemble gives a user a sufficient sense of "who I am" and "who you are." Social media profiles are a remarkably common technology for the contemporary production, modification, and curation of selfhood. With each carefully worded status update, we shape ourselves for presentation to a multitude of friends and followers, many of whom may be just casual acquaintances we once met at a dinner party, an art opening, or a scholarly conference.

In taking a longer historical perspective that builds out some of the back end of social media and other contemporary information technologies, I argue that the terms of our informational selfhood do not belong to social media, digital media, and other "new" media alone. Rather, what has proven powerful about social media is their success in leveraging a particular predefined composition of selfhood that was itself designed long before anyone ever dreamt of the internet, let alone the latest darling web app. I had a name, an address, a job, a degree, and specific identity characteristics before I joined any online social network. I was also well trained, as were my parents and grandparents, in presenting myself through these and other bits of informational identity. What social media adds to this long-familiar self-presentation is another iteration of a series of formats of our informational selves. Within a few short years we have all been retrained to present ourselves again through exactly the terms specified by the conventions of the newest social networks. But the fact of this retraining has itself become habitual.[18]

Our forms of informational selfhood are often criticized as abstract, reductive, artificial, and unrelated to who we truly are. But these objections rely on the conceit that there is something fundamental about who we are that is not itself always already the product of social and technological artifice. They rely, in other words, on the idea that there is an essential human nature, a real ontology of human being itself, that is somehow beyond the grasp of our informational dossier and yet within the reach of something else, perhaps a philosophical metaphysics.

My concerns here are different. I am interested in informational persons through their history, not their essence. In attending to historical conceptions of persons, my focus is not on what we *really* are but on what we *actually* are—a term that reminds us to attend impertinently to the *act* in the *actual*. Foucault once referred to his work as an analysis of "the pragmatics of the self."[19] The focus is on who we are in the sense of what we do—or, if not that, then who we can be in the sense of what we can do.

Among the many things we do today is database ourselves in a thousand ways. This is a feature of our selfhood insofar as it features in a range of actions peculiar to who we have become. Thus I agree with Schüll's description of "datafied subjectivity" as a "self in the loop": expressing a form of selfhood in which "digital tracking tools and their data can become part of the loop of reflexive recomposition" in such a way that we become constituted, and not merely mediated, by our data.[20] There is no essential self beneath all our data from which those data alienate us. Rather, on the Foucauldian view I share with Schüll, the pragmatics of the self are always coproduced with technologies of self. Our constitution by data technologies is only one instance of this. When we feel alienated by data, then, it is not because we are cut off from our essence, but rather because we have become estranged from other, prior, technologies of the self.

Data, therefore, is not the only thing about ourselves for which we need an account. I can readily acknowledge that we have both bodies and minds (I will even concede souls too, for those who want them). I have no interest in disputing such views. My argument is that these accounts, true as they might be, cannot possibly be the whole story about the peculiar forms of life in which we find ourselves today. There is something beyond the baselines of mindedness and embodiment (and ensoulment) that cannot, for all possible intents and purposes, be boiled down to them. In addition to being minded and embodied creatures, we are also our information. My argument is that our data are not mere externalia attached to us from which we might detach our truer selves as we please, but are rather constitutive parts of who we can be. Who we are is therefore deeply interactive with data. We are cyborgs who extend into our data.[21]

The questions I take up concern the manner in which data are important for the political and epistemic strategies through which we conduct ourselves and one another. Whether it is also the case that our data are fundamental to our nature—the essential ontology of our humanity—is a question I leave to the side. What we need, I propose, are philosophical perspectives for studying what is "important" that may not rise to the impressive level of the "fundamental." With an eye toward such a methodological proposal, I offer in place of a grand metaphysics of humankind a more humble genealogy of a specific kind of human subject.[22]

INFORMATION'S UNIVERSALIZABILITY

We informational persons have become our data. But our data has not defined us all in the same ways. In those differences lies a whole terrain of power and politics. Before developing an account of these operations of power—the main focus of the book—I need to establish the terrain on which I am suggesting we have all become situated. Who is the "we" who I am claiming have become our data? There are two answers to this important question. The book operates within, and across, the gap between them.

In one, relatively straightforward sense, I refer to everyone whose lives are conducted in important ways by information technologies. Taking this approach, it is an empirical question, albeit one of gargantuan proportion, to ask just how many people, or what percentage of people, are living in the shadow of their data. I suspect the number would be enormous, for the techniques through which data condition our lives define some of the basic practices through which we assert, claim, or are denied such infrastructures of living as citizenship, health, and education. One might even suspect that information in this sense invests literally everyone.

This brings me to a second, also relatively straightforward, sense in which I understand the "we" that is the concern of this book. This is the sense in which the "we" who have become our data refers quite literally to all persons today. Consider the difficulty of trying to live entirely outside of all possible data systems, in the style of the conjectural nightmare with which I began. And yet there are also wide swaths of persons who are deprived of some crucial bit of informational equipment to which most of the rest of us have access. These precisely are the political differentials of data about which this book makes an argument. But those differentials become politically and ethically salient only if we are all first situated on the same terrain with respect to informational technologies of identity.

One way to think of this second sense of how we have become our data is by way of two obvious objections. First, a humanitarian might object that my account ignores those who are paperless.[23] This, however, is the exception that proves the rule. Those who have been stripped of their standing (or who never received it) and made paperless, of course, inevitably find themselves demanding paperwork of some form or another. They are paper-less, not something-else-ful. We need an account of informational personhood in order to understand why being paperless is a political burden. Second, an anthropologist might object that such an account ignores some remote tribe that conducts its existence without any of the technologies we would think of as informational. This exception, too, proves the rule. Such a tribe would only be adduced in actuality as a counterexample to the imperatives of modernity and as such would stand as the anthropologist's purported exemplar of what is not modern. Perhaps the tribe is so unmodern as to successfully shirk data, but the anthropologist has (no doubt unwittingly) made the tribe into a data point for their argument, and thus along the way has given them informational identities like printed names. This is one sense in which we—all of us—cannot help but become our data today.

The universality of data's forays into subjectivity is largely a function of information's status as a universalizing technology. Information would have us accept that it is universal. It is difficult to resist its bid, for we all know how effortlessly we can see information anywhere and everywhere we can go. Any viable definition of information—that is, any definition that will not rule out any obvious instances of information[24]—will have to be so wide as to tend toward universal applicability.[25] Information is increasingly behind, underneath, and within all that we dream to do and all that we do in fact do.

We are therefore in a bind with information. We recognize that we ought to be skeptical about information's claim to universality. Yet we know we cannot simply refute information's success in making itself universalizable. If we find ourselves unable to both reject and affirm information's universality, then perhaps we ought to avail ourselves of some other option at our disposal. Rather than either denying the truth of information's universality or basking in the glow of that truth, we can ask how this truth came to be true. We can interrogate the genealogy of that which made information into such a powerful universal.

A genealogical approach helps us see how every actual universal was in need of mobilization at some point in its past. Genealogy, following Foucault, is a work of "rediscovering the connections, encounters, supports, blockages, plays of forces, strategies, and so on, that at a given moment establish what subsequently counts as being self-evident, universal, and necessary."[26] Thus, according to Foucault's description of his method, "instead of deducing concrete phenomena from universals, or instead of starting with universals as an obligatory grid of intelligibility for certain concrete practices, I would like to start with these concrete practices and, as it were, pass these universals through the grid of these practices."[27] This approach can be used to open up a crucial distinction between universality and universalizability, or what might be described as a distinction between eternal universals and historical universals, or in yet another way as necessary universality and contingent universalizability.[28] That which is simply universal is already everywhere; that which is universalizable can be mobilized to operate anywhere we want it to go—but it is crucial that this mobilization always requires the cooperation of numerous affiliates.

In the case of information, we can surely find it everywhere we go, but is this because it was already there, or because we have become so adept at both looking for it and leaving it lying around for others to find? Employing the distinction between universality and universalizability, the truly important question can be seen to be the one of how information came to be universalizable. Universalizability is what matters when information grasps us. The power of information is in its promise of, and success at, effective universalization. It is as a function of this universalizability that those who are in actual empirical fact defined by their data are able to embrace the idea that information has defined some of the terms by which everyone else today conducts themselves. The gap between the informational "we" that without a doubt will include every single reader of this book and the informational "we" that extends across everyone is that narrow margin between a universalizable technology being mobilized increasingly everywhere and some future iteration of that same technology that has already been successfully universalized.

Informational Politics

The argument that we have become informational persons, to be developed in full over the course of this book, involves two corollary arguments. One concerns the power that organizes a politics of informational persons. Another concerns the historical specificity of the emergence of this organizing power.

These two concerns are, of course, not unrelated. The historical problem is one I address by locating a historical moment in which we were becoming data by way of technologies that were still in the making and so still amenable to technical transformation and political intervention. From the perspective of our late present, this turbulent historical moment is one from which we can excavate operations and techniques of power that are now embedded deep inside of what we do. Seen from the other side, the political problem is one I address by historically locating the emergence of a politics of information in the work of forms and formats that all too often are presented as ahistorical in the sense of functioning perfectly well without regard for their history. These two arguments connecting past and power will be developed in full over the course of the book. But some additional orientation up front will prove helpful.

INFORMATION'S POWER

What form does universalizing information take when it is exercised on who we are and who we can be? What, in short, is distinctive about the form of power I call *infopower*?

A central argument of this book is that information is an exercise of power through the work of its varied and flexible *formats*. A paradigmatic instantiation of informational formatting are the checkbox-and-blank forms we all know so well. These are among the most ubiquitous of shapes into which we regularly squeeze ourselves. They exemplify how the work of a whole multitude of formats shape, constrain, and prepare whatever is collected, stored, processed, refined, retrieved, and redistributed as information. This formatting is rarely neutral. And yet it is regularly presented as such.

Infopower is exercised through this quotidian work of formatting. Through the formats that we habitually abide in our daily routines every time we fill out a form or check a box, be that on paper before a clerk or digitally on a terminal, we are enmeshed in the nets of infopower. This enmeshing involves a specific operation of power I refer to as *fastening*. Information's formats fasten us in a double sense that I leverage out of the two senses of this single word—formats both tie us down and speed us up. The fastening enacted in formats is an operation of power that canalizes and accelerates the persons it helps produce.

In telescopic summary, then, my argument across the book is that infopower as a distinctive modality of power deploys techniques of formatting to do its work of producing and refining informational persons who are subject to the operations of fastening. Even more concisely, technical formatting is an act of power that fastens its subjects to their data. By way of a comparison familiar to readers of Foucault (and to which I return at greater length in chapter 4), the normalizing operations and surveillance techniques of disciplinary power are analogous to the fastening operations and formatting techniques of infopower.

This account of the modality, techniques, and operations of infopower is so far only schematic. To provide a bit more legibility from the outset, I offer a preview of three kinds of fastening enacted by informational formats. I present this preview in recursive fashion by employing terms specified in an information technology standard that recurs throughout the book as a framing device. The standard is taken from a common technical definition used in information systems analysis according to which information handling can be parsed into three phases: data collection and storage (the *input* phase), data analysis and augmentation (the *processing* phase), and data dissemination and reproduction (the *output* phase).[29] In the input-process-output (or IPO) model of information, the three phases of information work sequentially from input to processing to output, and yet also form a loop or a circuit that sets enabling conditions for information to amplify itself.

Employing this standard model, my argument is that, at each moment of input, processing, and output, the work of information functions to pin down and speed up (or fasten) that which is being informationalized. Consider for a quick example the double fastening exercised by emblematic social media profiles (be they in brand-name behemoth networks like Facebook or narrow-segment spaces like Academia.edu). These profiles pin us down to prefab formats, categories, and conceptions that we then readily tie ourselves to. We proudly assert our interests in particular books, movies, and music by pulling them from suggestion databases populated by the interests of others (prompting a question about which obscure texts and albums have not found their way into that database). We can proclaim our friendship only with those who are also pinned down to the same profiles (who of your friends cannot be publicly displayed as one of your friends for the fact that they have not profiled themselves?). In addition to these pinning operations, social media also accelerates social recognition (e.g., friending) and interaction (e.g., sharing, messaging, and otherwise keeping in touch with the most distant family and the closest friends). Its work of quickening is evidenced by the steady tingle in our pockets produced by the constant stream of notifications social media send to our phones as an elicitation for ever more engagement.

A more detailed preview of these two sides of the power of fastening can now be seen by disaggregating social media into its three informational stages per my standard schema. The information we are able to *input* into social media in order to define ourselves is obviously limited (pinning us), but comes in user-friendly fashion (accelerating us). The varied informatics technologies (including but not limited to algorithms) that *process* that information provide us with standardized (tied down) suggestions about who to be friends with and what to buy, in a manner that makes it much easier to find that one person or one product among millions that we are searching for (sped up).[30] The social media profile that is *produced* on the basis of all this processed information looks more or less the same for us all (canalizing us) in such a way that we quickly learn to intuitively navigate one another (accelerating us).

While the au courant selfhood of social media exemplifies a contemporary form of fastening, it is hardly the most insidious site of contemporary infopolitics. Another poignant emblem can be located in the inferences drawn out of large data sets at the heart of algorithmic techniques common across corporate marketing and state security. Even those persons who manage to avoid all the major social networks will still be unable to avoid these databases for which none of us ever need sign up. As Louise Amoore details, even for those adept at the difficult work of keeping their personal information to themselves, "some element of their data will always already be in association with some other element gathered from other subjects and events."[31] It is today a fact about us all that we "will never be a stranger" to these algorithms and their stores of data.[32] These algorithms and databases also exemplify fastening in my double sense. The online consumers hailed with recommendations churned out by an algorithm have been carefully pinned down in an intricate taxonomy of categories so that they can be offered a shopping experience that quickens consumer behavior in the name of convenience. The airline passengers who are similarly marked by a security algorithm have also been microcategorized in such a way as to speed up not only their own processing according to security protocols but also that of their fellow passengers, who can now more quickly proceed to their flights after being cleared at the checkpoint.

When we are confronted with such contemporary contours of our informational selves, we may find ourselves remembering the time before the contemporary airport security apparatus, which was also the time before algorithmically personalized shopping, and also the time before ubiquitous social media. Some long nostalgically for whenever it was that we were not so rigorously categorized and scrutinized. Today's wondrous informational persons are, however, only the latest iteration in a long line of developments conditioning the informational person. The production of the subject of informatics has a long history. That history has now consolidated to a very dense point. If we can begin to understand the politics embedded in this history, then we will begin to gain a fuller sense of how to more actively reconfigure, remix, and even resist our informational selves. Opening up such possibilities of resistance requires attending to the specificities of how our informational personhood has been composed and recomposed over the course of its history.

An attention to such specificities of composition informs numerous recent contributions to the critical political theory of data, information, new media, the internet, the digital, computing, and other related focal objects. Many of these contributions, like mine, borrow extensively from Foucault's influential and innovative elaborations of power. Multiple surveyors have thereby located the power of data via Foucault's signature conceptualizations of power: these are the now-familiar arguments that contemporary informational apparatus are our newest instantiations of disciplinary power or biopower. In contrast to these arguments, my argument is that contemporary informational assemblies exhibit a modality of power not yet theorized in Foucault's work, nor in its wake. Informational persons, I argue, are being conducted by a power internal to information itself: infopower. This modality of power is neither a variant of nor reducible to biopower or discipline (this is a contrast I develop in detail in chapter 4). I thus do not aim to illuminate a data-driven biopower or a disciplinary informatics. I seek an inquiry into infopower as a mode of power in its own right.

My approach takes up a central provocation of Foucault's genealogies of power. Foucault's works exhibit, and thus motivate, a meticulous attention to the specificities of how power actually functions to produce, elicit, and sustain particular shapes of subjectivity.[33] According to these analyses, power is a matter of "the conduct of conduct."[34] It is about the shaping of who people can be in terms of how they can act—power concerns what we do to what others can do. Since people in different contexts can drastically differ from one another, we need to attend to the particularities of power's exercise. This attention to tiny differences of momentous consequence is part and parcel of Foucault's most provocative contribution to political philosophy: his argument that power does not always take the same form in the present that it has assumed in the past.[35] This is a crucial analytical insight that opens up for us today the possibility that we are in the midst of a distinctive mode of power when we find ourselves in the face of an assembly that orders, across a manifold of politico-epistemic practices, the informational persons we have become.[36]

Many insightful recent contributions to the critical theory of data are informed not only methodologically by Foucault's political theory from the 1970s, but also thematically by Donna Haraway's widely discussed essay "A Cyborg Manifesto" from 1985. One crucial argument of this piece is that the perils of the present are, at least in part, a function of "the informatics of domination."[37] If techniques of informatics produce us as informational persons, then they do so in ways that leave some of us dominated, and others of us dominating, even if unwittingly. Haraway's ideas call our attention to the need for a critical attention to information, and therefore also to the history of the politics of information. However, for Haraway, as well as for a majority of the critical theorists who have sought to reckon with the politics of information in her wake, the history of the informatics of domination has been largely confined to a residue of postwar "communications sciences and modern biologies" that all sought "the translation of the world into a problem of coding."[38] Just as I aim to redeploy Foucault's methodological approach to power, I also aim to reread, or reformat, such claims about the coding of the world in order to focus instead on an argument about the coding work that went in to making us into our data. In contrast to Haraway, and a substantial historiographical consensus following her intervention, I suggest that bringing the politics of information into view requires extending the scope of our historical analysis to the period preceding wartime information sciences and the postwar information theory to which they gave rise. I suggest, in short, that we need to attend to the work performed by information prior to its consolidation into information theory.

INFORMATION'S HISTORY

The digital personas, social media subjectivities, and surveillance-susceptible identities we have become are often described as entirely new kinds of creatures. New devices, new apps, and new media are advertised by their makers with the glitter of recentness all over them, such that we might even begin to suspect that they are somehow newly new, that newness itself has been reinvented as something other than what it once was. Not only consumers, but also critics, have followed the corporations in proclaiming the newness of all this informational accoutrement. In a widely cited book on big data, for example, Rob Kitchin concisely captured the buzz in speculating that "the data revolution is in its infancy."[39]

There is, however, a significant historical underpinning to our up-to-date information technologies. We begin to glimpse this history as soon as we find patience enough to look past the shimmering newness of our digitizing devices and look instead into the work these devices perform. A now-significant body of scholarship spanning a range of fields from new media theory to communications history to the history of science and technology has called into question the presumed avant-gardism of high-tech culture. These scholars have traced the terms of our informational culture back to earlier moments in a way that shows how social media, mass surveillance, and their informational underpinnings are anything but radically novel inventions of the past two or three decades. Yet within this scholarship there has also emerged a quiet but clear consensus about the historical origins of what Haraway called "the informatics of domination." The passing consensus holds that the terms defining our data-driven present were largely laid down in the middle years of the twentieth century, and more specifically right at the end of the Second World War, in the birth moment of what has since been named "information theory."[40]

The focal year for this consensus historiography is 1948. This was the year in which MIT mathematician-turned-metaphysician Norbert Wiener published his agenda-defining *Cybernetics: Or Control and Communication in the Animal and the Machine* and in which the MIT-trained mathematician and electrical engineer Claude Shannon published his equally ambitious "A Mathematical Theory of Communication" in the *Bell System Technical Journal*.[41] Wiener and Shannon, along with a host of other princes of science (they were almost all men), defined what soon came to be called the Wiener-Shannon theory of information.[42]

Were we to look into the milieu of 1948 beyond the mathematical complexity of the contributions of Wiener and Shannon, we would find there a growing obsession with information in a vast array of cultural sites. One familiar exemplar is George Orwell's dystopian novel of information and control, written in 1948 and published the following year.[43] *Nineteen Eighty-Four* completely immerses its plot and characters in a political reality whose primary motor is information. The success of the novel speaks to information not only as a feat of the scientists, but also much more widely as a cultural production that conditions how we think about ourselves and one another.

The consensus argument that 1948 gave wide berth to the birth of information culture is surely right, so far as it goes. But how far does this common sense actually take us? Surely it is striking that three of the most seminal texts of twentieth-century information culture were produced in 1948. But why was there such widespread obsession with information in that year such that it would have been possible for so many readers to want to take up information as a pressing concern? Why, in other words, were people ready for Wiener and Shannon and Orwell? Why their uptake? What were their conditions of acceptability?

In answering these questions, I argue that it would be a simplification to attribute the assembling of an information society to something as sparse as a theoretical program like information theory. What is needed instead is a complicating account of why so many people in so many domains found themselves so excited by information theory in that anxious moment at war's end. If Orit Halpern is right to observe that there was a postwar "dissemination of cybernetics and information theory throughout the social field" (and I think she is), then we need to explain what made this dissemination possible, rather than pursuing the typical route of attributing said dissemination to the supposed strength, efficiency, or profundity of these theories themselves.[44]

I argue in this book that the value of information conditioned the acceptance of information theory, and not the other way around. To make that argument, the periodization driving my inquiry departs from the consensus historiography account of postwar products grounded in wartime processes. I look instead to an earlier moment, one that was more turbulent in terms of its technical and social stability. This shift in periodization enables me to take seriously a series of questions that have been too much neglected in the recent critical historiography of information. What was information before information theory? What was the status of information before it came to be theoretically consolidated in the span of a single year, as if all at once, by two mathematicians in writings that did not contain the word *information* in their title but rather the word *communication*? If information theory were the originating moment of contemporary information cultures, then how would we account for a widespread and self-conscious reckoning with information before those origins? What kinds of practices were Wiener (b. 1894), Shannon (b. 1916), and their contemporaries acculturated into? What kinds of social practices were people then immersed in, such that they could be so concerned with information as to elevate the technical formulae of mathematicians into perceived keys for steering a course through the next century? What made possible that gleeful uptake of glittery cybernetics in the 1950s so that James Baldwin would later remember it as "the cybernetics craze"?[45] What made possible the more lasting enrollment of information across a widening cultural panorama such that, by the 1970s, terms like *information society* and *information age* had become part of the lingua franca?

Investigating these questions requires turning to moments prior to the consolidation of information theory. It requires excavating an ensemble of information technologies and practices of informatics that helped make information into something that appeared to stand in need of a theorization capable of enforcing a consolidation. The first decades of the twentieth century are of interest, then, because they were a time when data was on the cusp of consolidation. Neurath's Isotype project offers a convenient exemplar. The project was progeny of the 1920s and 1930s. By the time Neurath was composing his autobiography in the 1940s, Isotype was already institutionalized and formalized. I focus on moments in which information technologies like Isotype were in formation, rather than on the subsequent proliferation of what was by then already formed. Information just prior to the consolidation of information theory was on the verge of freezing—one degree of difference from its zero point. I am interested in this cusp of consolidation because it makes visible two tendencies of our informational personhood in conflicted contemporaneity: the transformability of persons who were becoming their data, and the consolidation of persons such that they were increasingly obliged to become their data.

Although I depart from the focus of the historiographical consensus, I do not intend to refute it. Indeed, my argument relies on the consensus view on some crucial points. I do not here track forward the emergent informational person of the early twentieth century through adolescence and maturation in subsequent decades (there are no chapter-length studies of the late 1940s, the 1970s, or the early 2000s). I do not rehearse work that has already tracked these movements, because I agree with the consensus view that the postwar period was indeed a moment of remarkable uptake of information theory that in and of itself surely explains something about our present information cultures. Thus my point of dissent is just this: the uptake of information theory that helps explain our present is itself also in need of explanation.

There is a more general commitment that is operative here: there are no un-explained explainers. This commitment is crucial in dealing with universals like information. As Gilles Deleuze and Félix Guattari observe, "We think the universal explains, whereas it is what must be explained."[46]

The ensemble of information practices I excavate from the early twentieth century in an attempt to explain the success of the universalizing information theory that followed it is at once a cultural, medial, scientific, and political assembly. In attending to the functional work of information in otherwise disparate domains from the 1910s to the 1930s we can begin to make out the emergence of what I call "the informational person": the person who, like you and me today, is defined by the data sets around which conduct accumulates. That person, the informational person who devotes so much energy to curating the information composing his or her life, is in some crucial ways a creature of the 1910s to the 1930s. One hundred years later we are only just now beginning to bring into view the politics specific to that creature that we have all become. Bringing that politics into view, and resisting the misuse of our informational selves, is likely to be a matter that will involve confronting the malleability of designs that we take to be unchangeable, the transformability of arrangements that we take to be consolidated, and the recoding of technologies of the self that we have unthinkingly allowed others to code for us.

Design and Method

How We Became Our Data attends to the role of informatics in the production of shapes of selfhood, or modes of subjectivation, that eventually helped make information theory seem so attractive and then afterward helped make the information society seem such a plausible explanation. The argument can only unfold in the details below, but it will be helpful to orient those details with a brief summary of the arc in which they are embedded. Doing so also occasions consideration of methodological questions that further orient the reader. After a summary description of the book's general design, three such methodological questions are brought to the fore. How do I methodologically mobilize a critical inquiry into the history of informational personhood? On what materials is my critical inquiry trained? What are the boundaries that frame these objects of inquiry?

DESIGN (OR, PLAN OF THE BOOK)

Part I of the book tracks early moments of the informationalization of selfhood across a series of forms of subjectivity. Chapters 1 through 3 thus genealogically analyze three domains of identity that index significant paradigms of our ongoing infopolitical present: documentary identity (chapter 1), psychological identity (chapter 2), and racial identity (chapter 3). The first chapter tracks emergent informational persons in the contexts of the bureaucratizing paperwork of the standardized birth certificate. The second traces personality metrics amid an episode in the stabilization of a prestigious subfield in scientific psychology. The third traverses the informatics of race in the context of real estate appraisal so as to excavate the informational mechanisms buried within the history of American home loan practices like redlining. With respect to each of these domains, there was a crucial shift between the mid-1910s and mid-1930s which yielded the informationalization of aspects of identity that still remain with us one hundred years later. In each field I locate the operation of a politics of information. Put in the briefest terms, the chapters analyze the datafication of life, mind, and body (more specifically, birth, personality, and race).

The particular mélange here under survey prompts an immediate question. Why these three vectors of informationalization? What could possibly hold them together? A not-unreasonable charge of eclecticism is here in the offing. I offer a few preemptive responses.

One, already noted, is that the three vectors of documentary, psychological, and racial identity on which I focus are crucial for who we can be in the present. The possibilities for, and limitations of, our selves are today deeply informed by bureaucratic paperwork, psychology, and race. These are not the only aspects of our identity that matter to who we are, but they clearly do matter, and a great deal. They matter so much, in fact, that they can limit us unevenly.

A second response is that these three cases allow me to focus on a broad range of institutional actors, including statecraft (in the case of identifying documents), science (in the case of personality psychology), and capital (in the case of racialized appraisal). This is no small advantage, though it is not a particularly explicit feature of the ensuing analysis.

A third response is that my topics are reflective of contemporary concerns with data infrastructures. My hope is that an articulation of these histories of datafication can speak to an ongoing cultural conversation on topics such as data breaches and identity theft, debates over the ethics of artificial intelligence, and analyses of algorithmic bias and discrimination. Interest in these issues is of increasing gravity today—these are topics that have been the subject of at least occasional (and in some cases frequent) front-page reporting in recent years. Perhaps there is such great interest in these matters because they represent three great promises of our contemporary informational milieu. One is the promise of complete surveillance and inescapable data capture. Another is that of the computability of the deepest interiorities of human mental and emotional life. And a third is the utopic dream of a society that would transcend divisive racial and ethnic differences, captured so well in all its ambiguity and ambivalence by that old 1993 cartoon in the *New Yorker* in which one dog sitting at a computer terminal says to another on the floor, "On the Internet, nobody knows you're a dog." Each of these promises is deeply felt, but also profoundly ambivalent. For just as they are exciting dreams for the many who pursue them, they are at the same time threats to so many who would become the subjects of such programs. What a history of documentary identity, personality psychology, and racialized datafication helps us see is that such promises and perils, despite seeming so new, are in fact the latest iteration of much longer histories, and ones that moreover were fraught from the start.

In sum, then, my three chapters feel so disparate because I need them to. Were they to give the appearance of a small number of puzzle pieces that seamlessly fit together, then my argument would be restricted to just that seamlessness. What I am interested in, however, is the ranging universalizability of information. Attentive to the contingencies by which our data increasingly invests multiple domains of identity, I need no theme other than information as the thread that runs through the fields I consider. A real disparateness of cases (rather than a tidy coherence of examples) is precisely what is of value in excavating how information began to run pell-mell through so many such fields, not as a secret necessity, but rather as an historical contingency.

If that is a sufficient reply to the charge of eclecticism—and I hope it is—then another question that is prompted in turn concerns how one might navigate such a disparate selection of topics. To orient this multiplicity, I employ the above-noted technical standard from information management: the IPO model. This standard schematizes my analysis. But it is a schema only. Each field into which I venture below relied on all three phases of information registered in the schematic model: inputting, processing, outputting. Yet in each there is also a particular kind of information technology that appears most prominent. In the consolidation of identifying documents of administrative bureaucracy, it was the *inputs* that were prominent: the technical achievement in this case had to do with matters establishing successful data input practices. In the measure of personality traits in psychology, it was the *processing* algorithms that were the leading information technology helping personality psychology secure its status against competitor programs. In the datafication of race for real estate accounting, what appears most prominent, especially at first, are the *outputs* (in the form of valuation surveys and property maps, including the infamous redlining maps) that were massively effective in disseminating subtle techniques of racial segregation. The privilege I offer to a single dimension of information in each chapter is not contrary to all three dimensions being necessarily copresent in each according to the technical standard itself.

After a sustained survey of three episodes in which information technologies of formatting were applied to persons, part II turns to critical implications, particularly with respect to contemporary problematics in political theory. Two such problematics are central.

The primary contribution of chapter 4, as briefly noted above, is that the formats and fastenings of infopower are irreducible to the politics of discipline, biopolitics, and state sovereignty that are familiar from the work of Foucault and other recent critical theorists. Though infopolitical fastening often hooks up with earlier-developed political technologies of normalization and regulation, technologies of canalizing and accelerating are in need of a diagnosis in their own right that does not misread them through concepts appropriate to what are in actuality quite other operations of power. The political stakes of such a recalibration of our critical diagnostics matters not only for understanding the history of who we have become, but also for confronting possibilities of who we might yet be.

This leads to a second argument in political theory, taken up in chapter 5. In the same way that infopower is irreducible to other modes of power, so too the resistance calibrated to infopower is irreducible to mainstream theories of democratic deliberation that presuppose information in such a way that they cannot confront it as a political problematic in its own right. I have in view here those influential communicative accounts of democracy that have structured much of normative political theory for the past few decades. My primary interlocutors in my assessment of communicative democracy are, as with my engagement with Foucault, sources of methodological insight from whom I draw throughout the book. Nevertheless, I argue, Jürgen Habermas's and John Dewey's theories of communicative democracy are structurally unable to confront information itself as a political problem. Rather than suspending communication-centered politics by way of a turn to aesthetics (a recourse so prominent in contemporary political theory), I turn instead to technics. On this view, I suggest, resistance to infopolitical fastening is best mounted at the level of designs, protocols, audits, and other forms of formats.

METHOD (OR, ANALYTICS OF INQUIRY)

Having asserted that the argument of this book involves pressing beyond Foucault's conceptualizations of the powers of disciplinary normalization and biopolitical regulation, I need to be clear that the manner in which this argument is made is deeply indebted to Foucault's genealogical method. To clarify this, I offer a distinction between Foucault's concepts and his methods. The method at work in this book is heavily indebted to Foucault's genealogical analytics of his present.[47] I have sought to leverage that method beyond Foucault's own present so as to conceptualize our contemporary politics of information.[48] Foucault himself could not have employed his genealogy for the purposes of conceptualizing our "narrow now" of data.[49] Social media profiles, big-data analytics, and digitalized state surveillance are our task. Foucault's present still applies to ours, but where we are today has also clearly gone beyond the limits that Foucault would have been able to discern in his time. It is in this sense that I mobilize Foucault's analytical strategies (i.e., genealogical method) without getting tangled up in his concepts (e.g., discipline and biopower).

In appropriating from Foucault his "method," I want to underscore that I do not intend the term in any ponderous sense. A method is not a guaranteed way of achieving a goal; it is not a foolproof recipe; least of all is it a surefire algorithm. A method is simply a way of proceeding. It is, as per the Greek *methodos*, a mode of traveling. I here seek to travel, to move, and to mobilize acts of inquiry amid a series of objects of study. So that we do not become lost when we inquire beyond our armchairs, we need ways of proceeding.

What, then, is characteristic of genealogy as a method for inquiry? Allow me to attempt to boil down to a single paragraph what at one time it took me an entire book to work out.[50] Genealogy is oriented by three philosophical commitments. First, genealogy is a practice of critique. This does not mean that it stands in judgment of that which it surveys or tells us what is wrong with the world. Rather, genealogy is critical in that it explores the limits of what we can do in the present. These limits may be judged to be good or bad; but genealogy is concerned with the conditions of possibility that define the present in such a way that certain actions simply are not possible for us. Second, genealogy is concerned with conditions of possibility insofar as these conditions are contingent, rather than necessary. In this, genealogy requires that philosophy involves itself in history, which is to say that genealogy affirms an internal connection between philosophy and history. History without philosophy would not know where to look—and philosophy without history would have nothing to see. For it is history that deals with contingencies. History tells us how things just so happened to have happened. In focusing on conditions that just so happen to constrain our possibilities for action in the present, genealogy makes no claim that everything is contingent (or that nothing is necessary), but rather only focuses our attention on that which is contingent. Of course, the really interesting cases are those in which we thought we were constrained by the force of necessity, but where it turns out that the conditioning constraints are only contingent and yet also no less constraining for that reason. Third, genealogy is committed to complexity. Correlative with an emphasis on contingency, this commitment involves the rejection of simplifying explanations that would deduce from some single cause the conditions bearing on the present. That which conditions us contingently does so in part because it is always the effect of a compromise between a multitude of forces in struggle with one another. Such compromises are never simple, never easy, and never innocent. The limits of who we can be, and the boundaries that we find that we must remain within, are always the result of partial settlements that may come to be unsettled again.

While my method here is located largely in the wake of genealogy as modeled by Foucault, the philosophical inspirations motivating this work are many. Alongside genealogy, I also take up perspectives indebted to philosophical pragmatism (including not only the classical pragmatisms of Dewey and William James but also more recent variants of pragmatist critical theory like Habermas's), and Latourian actor-network theory, both of which counsel to always keep in view the functional work, or transitional turnings, performed by things in the way that they act.[51] A common analytical thread runs through genealogy, pragmatism, and actor-network theory: a vigilant attention to conduct.

Conduct is a privileged site of critique because who we are is a function of what we do. With this, I offer an account of how a mode of power came to define us as particular kinds of persons, that is, as particular kinds of actors whose possibilities for action are conditioned in particular ways. The attention to conduct central to this account is oriented primarily, though not entirely, by Foucault's idea of "modes of subjectivation," which functions as a compass for "a history of the different modes by which . . . human beings are made subjects."[52] An analytics of subjectivation brings into view a crucial question. What are we made to do? I here give that question its double sense whereby it points both to the imperatives of power that guide our conduct and also to the histories whereby these imperatives were crafted.

OBJECTS (OR, SOURCES UNDER SURVEY)

How did informational people like you and I become ubiquitous when so many people just one century ago (and nearly all of them three centuries ago) would hardly have recognized who we were becoming? To understand the specificities of how this happened, I focus on the particular techniques by which informational personhood was made to operate. This style of analysis raises a number of crucial methodological and historiographical problems.

One concerns source materials. The sources from which my analytics of techniques has been assembled are not the prosaic texts that critical disciplines like philosophy are often trained on, and only occasionally the argumentative work of treatises, monographs, and articles. My analysis relies much more extensively on a different kind of paperwork: printed blank forms, the protocols used to process them, and the reports and displays that they were used to produce. Even where I am attentive to standard prose (in my discussion of the articles and monographs in which personality psychology gained scientific status, for instance), what is also of interest in these documents are the modes of information display they enact (charts, graphs, tables) and the underlying informatics (formats, databases, algorithms) on which their prosaic arguments depend. Across the book, it is not just what documents say that is of interest, then, but also what all these many different kinds of documents are doing.

This raises a question concerning how such documents are to be read. The very work of attending to the kinds of sources I do already implicates my analysis in an approach that departs from the hermeneutics of meaning. I emphasize instead the pragmatics of functionality. If this seems like a strange accommodation to make (and it will to some readers), consider critical media theorist Tiziana Terranova's claim in her "Communication beyond Meaning" that we find ourselves today in the midst of "contemporary cultures where informational dynamics are increasingly gaining priority over the formation of meanings."[53] As Terranova explains: "It is not so much a question of meanings that are encoded and decoded in texts but a question of inclusion and exclusion, connection and disconnection, of informational warfare, and new forms of knowledge and power . . . not so much the play of meaning but the overall dynamics of an open informational milieu."[54] According to this perspective, making sense of contemporary informational dynamics requires of us analytical modes that do not presume at the outset that these dynamics are always grounded in meanings that await intellectual acts of hermeneutic deciphering. Experimenting with an alternative approach, this book is focused on detailing specific families of techniques of information collection, analysis, and distribution. The shift from a hermeneutics-first style of philosophy to a pragmatics-centered analytics is, again, correlative with the shift from prosaic archival sources to a more motley assembly of documents that say very little but that format a great deal: birth certificate blanks, housing appraisal forms, and personality test questionnaires. This shift in source material clarifies why the book's analysis cannot be focused on the arguments that great theorists put forward on behalf of techniques of informatics (nor on the arguments that critics may have given against them), but rather must take as its focal objects the technical operations of information technologies and the humble technicians who put them into motion.

In this respect, I again stalk Foucault's practice of critique as an analytics of the pragmatics of techniques. Consider how replete is his *Discipline and Punish* with techniques and technologies. The focus is on "techniques that serve as weapons, relays, communication routes and supports for the power and knowledge relations that invest human bodies."[55] Notably, his analysis is not an attempt to decipher the arguments that so many Benthams may have given for so many Panopticons, but is instead an exposition of how diagrams like that of the Panopticon were made effective, actual, and practical. Consider also how Foucault's *The Birth of the Clinic* suspends the assumption that inquiry must "uncover that deeper meaning of speech that enables it to achieve an identity with itself, supposedly nearer to its essential truth" by instead favoring a mode of analysis that "suppos[es] no remainder, nothing in excess of what has been said, but only the fact of its historical appearance."[56] Ian Hacking is right to claim of Foucault that his approach "is the very opposite of hermeneutics."[57]

Historian Arlette Farge, a collaborator of Foucault's, offers an apt description of this feature of genealogy.[58] To conduct the analysis of discourses, she says, "we must therefore look deeper, beyond their immediate meaning," so as to apprehend "the heart of the circumstances that permitted and produced them," or what she later calls "the conditions of their appearance."[59] This form of analysis also appears, after Foucault's prompting, in recent media theory, specifically in the media archaeologies of Friedrich Kittler.[60] Though Kittler's work forms a foil for my own here in terms of the contents of some of his historical claims (most notably his thematic obsession with war and his historiographical focus on 1948), the methodology informing his work is crucially important for my pursuits. What stands out most in Kittler is his post-hermeneutic attention to technical details: standards, schematics, and, as I would have it here, formats. Kittler describes his approach as yielding "discourses on discourse channel conditions."[61] The conditions of appearance to which Farge invites us to attend are for Kittler to be construed as technical conditions of appearance. What, Kittler's books ask, are the technical channels through which discourse can be made to flow? Or, as I ask in this book, what are the technical formats through which our subjectivities have been remade? Thus is my inquiry one possible subspecies of Kittler's provocative claim that "where thinking must stop, blueprints, schematics, and industrial standards begin."[62] Where thinking must stop, we can either endlessly decipher the meaning of thought's finitude, or we can pivot toward the histories of the chan-nel conditions that thought has bumped up against. Kittler's pivot, as Caroline Bassett notes, "sets a (technological) cat amongst (hermeneutical) pigeons"; it is a discourse on channel conditions for which "not only representation and language, but also human experience and human interpretation (meaning-making), are . . . misguided."[63] This is not to insist that meaning is always be-side the point (what could that even mean?), but only that meaning is not always to the point. Following Kittler and Farge following Foucault, then, my mobilization of genealogical methodology is one that would be postinterpre-tivist in enacting not an analytics of meaning, but rather one of functionality, action, and event. Where Foucault wrote of stone structures and clinical tech-niques, I attend to the formatting work of printed blanks, questionnaire tests, and template forms.

SCALES (OR, FIELDS IN FOCUS)

No genealogy can pretend to be an exhaustive history of its subject matter. This is all the more true for genealogies of purported universals like infor-mation. There are always limitations of topic, time, and space.

Topically, it follows already from my argument that there can be no such thing as an exhaustive accounting of the history of the production of the in-formational self. Largely left to the side in what follows are a range of other domains in which information has come to be a defining condition. There is, for instance, an economic informatics in life insurance, in systematic man-agement, and more generally in corporate capitalism and its afterlife in fi-nance capital.[64] Or, in a quite different domain, there is an informatics of life itself in a history of genetics that consolidated in cracking the genetic "code" that is the information-processing machine for who we are.[65] In the face of such a rich array of interesting sites of investigation, why do I decline to con-duct an exhaustive analysis? Quite simply, because I must. This is precisely what the universalizability of information ensures. There can be no exhaus-tive history of a universal—there can only be selective histories of how any universal came into possession of its universalizability. To track every last blessed instance of informational personhood would be to track all of us all of the time. Even the most powerful algorithm could not be made to do that.

With respect to limitations of historical periodization, I have noted that I do not here focus my analysis of the informational person on the mid-twentieth-century moment familiar from contributions to the consensus his-toriography. The reasons in favor of my earlier periodization were already noted above. But what about the limitations on the other side of my period?

In seeking to historicize beyond the limits of the consensus historiography, do I really historicize enough? If my argument is that we need to move the historical focus to a moment that is prior to postwar information theory, then what about those moments prior to my own?

My analysis is focused where it is precisely because this is the moment in which the universalization of an informatics of personhood began to consolidate. This moment was a cusp of consolidation that bears witness to multiplicitous conditions (hence my motley of subjects across my three chapters) rather than a singular origin.[66] One could always find antecedents to the information technologies I here discuss. These are not my focus, however, because with respect to the fields I here survey, these antecedents do not so much demonstrate the stabilization of universalizable technologies as they demonstrate prototype technologies.

Consider, for instance, how Daniel Headrick's *When Information Came of Age* offers an alternative historiography in its narration of a period from the eighteenth to early nineteenth centuries in which a range of technologies were developed for the organization, transformation, display, storage, and dissemination of information.[67] I depart from this periodization for the reason that information technologies of that time were rarely successful in achieving universal applicability. Historian Gérard Noiriel has investigated an 1820 project by the French Ministry of Justice surveying the enactment of a 1792 law requiring public officials to keep registers of vital events (births, marriages, and deaths) for all inhabitants.[68] He concludes, "There was nowhere that the registers were being correctly maintained."[69] Political scientist James Scott notes about another such project: "In 1791, the Revolutionary State in France required all prefectures to furnish the 'name, age, birthplace, residence, profession, and other means of subsistence of all citizens living in its territory.' Only *three* of 36,000 communes replied!"[70]

Schemes of informational identity in these earlier centuries tended more toward future visions than present actualities. Ian Hacking describes a 1685 proposal by G. W. Leibniz for a fifty-six-category evaluation of a state and its population: "Like so many of Leibniz's schemes, such a tabulation was futurology that has long since become routine fact."[71] It matters, Hacking argues, precisely when futurology becomes facticity. Consider two other envisioned informatics projects that complement Hacking's Leibniz in that they too are dreams from since-canonized philosophers. Jeremy Bentham and J. G. Fichte are better known for more-mainstream contributions to political and moral theory, but their policy proposals for informational identification are equally interesting, if only because they were so much ahead of their time, even if not quite ahead as Leibniz.[72]

Writing in the 1770s, Bentham stated that the fundamental question at issue in identification is this: "Who are you, with whom I have to deal?"[73] Such a query poses innumerable problems, Bentham noted, given the many manners of deception with which it might be met in reply. "It is to be regretted," he observed, "that the proper names of individuals are upon so irregular a footing."[74] Facing such common problems as two persons sharing a single name, Bentham advocated "a new nomenclature . . . so arranged, that, in a whole nation, every individual should have a proper name, which should belong to him alone."[75] He passed over in silence the organization of the massive governmental task of administering the registration of unique names, though he did admit that his proposal might produce much administrative "embarrassment."[76] Bentham also proposed something much more radical—namely, repurposing "a common custom among English sailors, of printing their family and christian names upon their wrists, in well-formed and indelible characters."[77] His proposal, in other words, was that of identification tattoos, which, he argued, "should become universal."[78] Bentham, famous as the visionary of the modern surveillance *dispositif* architected in his Panopticon, also anticipated contemporary information politics. The Panopticon was meant primarily for prisoners, and perhaps also for some classes of workers, but the identifying tattoo was intended for us all, and indeed could only possibly work if applied in universal fashion. "Who are you, with whom I have to deal? The answer to this important question would no longer be liable to evasion."[79]

Writing two decades later, the German idealist Fichte took up the same problematic of identification as an aspect of the rights of "the police—its essence, duties, and limits."[80] He was unambiguous about the importance of identification for police work: "The principal maxim of every well-constituted police power must be the following: *every citizen must be readily identifiable, wherever necessary, as this or that particular person.*"[81] If Bentham's solution to this problem feels somewhat fantastic, then Fichte's seems positively mundane: "Everyone must always carry an identity card with him, issued by the nearest authority and containing a precise description of his person; this applies to everyone, regardless of class or rank. Since merely verbal descriptions of a person always remain ambiguous, it might be good if important persons (who therefore can afford it as well) were to carry accurate portraits in their identity cards, rather than descriptions."[82]

What is of interest in these two late eighteenth-century cases is the appearance of proposals of universal identification that remained for more than a century just that: proposals. Gargantuan projects of universal identity registration at the fine-grained level of individuality could be imagined at the end of the eighteenth century, and even debated over the course of the nineteenth, but they would not be operationalized in anything approaching a universal fashion until the first decades of the twentieth. Eventually we became just who Bentham and Fichte hoped we would become. Our data today is a part of who we are. Our names may not be inked onto our flesh, but we are for pragmatic purposes tattooed by an informational identity that we also regularly carry on our persons in the form of identity cards. Our now-flourishing projects of information as a technical machinery for manufacturing identity have a deep legacy in modernity. They are realizations of old dreams.

It is no surprise, then, that there are numerous instances of earlier information within each of the fields I survey in the book. With respect to the identification documents discussed in chapter 1, one might locate earlier origins in passport papers dating as far back as the 1400s.[83] With respect to the discussion of personality measures in chapter 2, one might look to earlier instances of mental and physical measure, or even to first-generation anthropometric sciences like phrenology and craniometry.[84] And with respect to the discussion of what I call the "informatics of race" in chapter 3, one antecedent to my periodization would be to take up the role of records, paper, and data in the maintenance of slavery, as for instance in Stanley Engerman's study of the role of registration in the international slave trade,[85] or more recently in Simone Browne's book *Dark Matters: The Surveillance of Blackness*, a text to which I return in that chapter.[86]

In each of my domains, histories of antecedent technologies can take us further back in time, but the cost will be that of a diminishing scale on which those technologies operated. This book places the historiographical emphasis on universalizability. Much comes into view in focusing on earliest instantiations, but so too does much else come into view in focusing on scalable universalization.

The periodization here under survey thus differs from both the consensus historiography of information theory that is focused on the postwar years and the multiple histories of the origins of information technologies that push the historical envelope back into earlier moments in modernity. The slim few decades that have fallen between the cracks of historiographies of the postwar information explosions and those of the floods of printed matter in earlier centuries are a relatively neglected era in the historiography of information (and as well, I might add, in the historiography of subjectivity). Widely neglected, these decades are, however, the subject of a growing number of recent studies of information technology.[87] I thus here ask, alongside James Purdon's history of informatics and literature in the early twentieth century, *Modernist Informatics*: "What cultural pressures went into the making of information as a concept before it was conscripted as a theory?"[88] To the small handful of studies sharing Purdon's question I seek to contribute a genealogy of a generalizable class of informational persons. My contribution thus aims at a wider target than that yielded by Purdon's analysis of a "new kind of informatic identity" in literature and Dan Bouk's history of the "statistical individual" and "paper people" of early-century life insurance practices.[89] What I am interested in is precisely how these and other narrow domains where informational personhood was present could have put into operation technologies that were mobilized to become universals.

These issues of temporal scale raise a final series of questions concerning the spatial limits of my study. Why do I limit the birthplace of my universal informational persons to a small handful of American vistas? What do I lose by not attending to other locales? Not nothing, to be sure. Having noted how early nineteenth-century French dreams for a meticulous accounting of the population remained largely dreams, why do I nonetheless take up neither standardized birth registration in France nor the English parish registries, both of which preceded American birth registration standardization?[90] Why not consider colonial practices of identification, such as British fingerprinting in India in the late nineteenth century?[91] A sufficient answer to these questions can only be developed over the course of the book, where my argument concerns how a whole raft of quite different practices all began to invest subjects with data in the early twentieth century. For now, the short answer is that I restrict myself to touring the American scene because it was a major province within which was engineered much (though certainly not all) of what operates today as our contemporary universal informatics of persons. I do not mean only that America has been the epicenter of our contemporary data technologies. Nor do I mean only that information has become the chief global export of American culture in addition to being one of its most distinctive contributions to its own cultural identity. Both claims may well be true, but they are not what is decisive.

What matters, rather, is that it is because of their massive informational exports that American corporations, governmental agencies, universities, and even social critics have also managed to export a very parochial ideal of universality. America has been the province within which was worked out the terms of so many of the most scalable implementations of universalizable information technologies, including so many of those information technologies that increasingly define who you and I can be. As Tung-Hui Hu remarks of our contemporary networking technologies, they were engineered in a way that "reflects a universalist world view that tracked closely with American political ideals."[92] So too, I argue, of some other of our technologies of information that, many decades prior to network protocols, had already begun the work of a formatting that promised a particular instantiation of universality. Although the historiography that follows is a decidedly American one, it is such because it is concerned with a time in which America came to regard itself as a world-universal project. For better or for worse, in some instances, like that of the internet, the United States pretty well succeeded in its grandiose ambition of universalizing some of its most parochial ideas of universality.[93]

What I am suggesting is that those whose lives are increasingly formatted by information technologies are in the main being fastened by techniques that were in many instances designed as reflections of a provincial ideal of universality. If that is right, then paying attention to a privileged province as the scene of emergence of a universal is useful for understanding how information technologies were designed such that they could circulate as universalizable. The questions this book asks, then, are about how a very particular and powerful form of universalizability came to be contingently composed in a series of political technologies of data that format much of what we do.

PART I

Histories of Information

"Human Bookkeeping":
The Informatics of Documentary Identity, 1913–1937

Who Are You?

Jemma Cook was born in a car on the way to a Los Angeles hospital on a summer day in 2017. Jemma's mother, writer and editor Jia-Rui Cook, described giving birth in a car as complicated in all kinds of ways. There was, of course, the biological mess of blood and birth water all over the passenger seat. But, as Cook relays in the *New York Times*, "the biggest problem, it turned out, was one we would never have foreseen: getting a birth certificate for our baby."[1]

Jemma's parents were without delay in taking her to the hospital for natal care (and delivery of the placenta). In some states, hospitals have a duty to provide access to a birth-registration system in virtue of being the first to provide medical care to the infant. But California law requires that hospitals assist in birth registration only for infants born on their premises.[2]

Jemma's parents were thus made to turn to the local agency charged with administering birth registration. It took them over a month to get an appointment with the Los Angeles County Department of Public Health (LACDPH). At the appointment they were required to produce five pieces of evidence that together would be sufficient for establishing Jemma Cook's identity and her impending lifelong answer to that first of bureaucratic questions, "Who are you?" According to the LACDPH website, the Cooks were required to produce: identification documents for both parents, proof of Jia-Rui's pregnancy, proof that Jemma was born alive, proof that the birth occurred in Los Angeles County, and proof of a witness.[3] Proof that a mother was in Los Angeles County during birth can be established as easily as producing a utility bill for the month of birth. The identity of the proving witness can be easily established by their accompanying the mother to the LACDPH. A little less straightforward, however, is proof of a live birth where no licensed doctor or midwife was present to certify the fact. In Los Angeles County, proof that a

baby has been born alive can be established, somewhat remarkably, by bringing the baby being registered to its LACDPH registration appointment.

This specific bureaucratic requirement expresses the circularity haunting
all proof of identification. Proof of who we are is often needed. But how is
that proof ever established in the first place? How can the first bit of proof of
identity be put into place? Only by a silent moment of conjuration in which
we all knowingly blink at the miracle of something being created out of almost nothing at all. The creation of identity is strangely akin to that other
miraculous moment of a conception whose eventuality is a baby come crying
into the world.

The broader reach of this conjuration act of officialdom is expressed in
a small snippet of Jemma's birth certificate accompanying Cook's article. In
a narrow image whose background is official yellow with blocky black type,
Jemma's birth certificate lists her "Place of Birth" as "Automobile." Never mind
that "Automobile" can never be a place (though it is of course possible for "an
automobile" or "this automobile" to be a place in certain circumstances). For
what is remarkable here, and what we usually pass over in one of our silent
blinking moments, is the form's request for "Place of Birth—Name of Hospital or Facility." Therein does the birth-registration record quietly format
what can even count as the place where one has been born. Just as a wiggling
baby who is brought to a county health agency can stand as a kind of proof
of the actual vitality of some identity, so too can "Automobile" become an
actual place. In both cases, this is made possible—even necessary—because
the form requires it.

It is a trivial observation but a remarkable fact that there is a format to the
record of being born. What those formats are, their specific design and the requirements they solicit into being, come to have enormous importance to who
we are, and who we can be allowed to be.

What Is Your Name?

Consider another ubiquitous technology of informational identity: the name.
Names have become such a routine part of our identity that we easily forget
that, in order to become effective parts of who we are, they must be made to
fit the requirements of identification they are to be used for. Fitting names to
the constraints of modern governance was anything but the simple task we
might take it for whenever we effortlessly relay our name to a cop or clerk
today.

According to historian Jane Caplan, the technology of the name as a universal descriptive designation did not become usable for identification in

much of Europe until as late as the nineteenth century.[4] In 1878 a lawyer in England noted that the country had no laws governing names.[5] In many countries, the situation on the ground was even worse than that on the books. In Scotland in 1852, the registrar general reported that surnames could "scarcely be said to be adopted among the lower classes in the wilder districts"[6] while ten years earlier one official reported that the small town of Buckie was home to no fewer than twenty-five individuals by the name of George Cowies.[7] In France the mayor of Metz noted at the late date of 1908 that it was quite common "to find a son bearing a different surname from his father, a brother from a brother, and to discover individuals having borne one name decide to take another."[8] In 1885 an attorney corresponding with the United States Department of State sought to defend a client who wished to spell his name differently on different documents.[9]

Assessing the relative recentness of the reliable standardization of names, political theorist James C. Scott concludes that "the invention of permanent, inherited patronyms was . . . the last step in establishing the necessary preconditions of modern statecraft."[10] According to his classic *Seeing Like a State*, large organizations at the heart of bureaucratic modernity (most obviously states) need to be able to "manipulate" the people with whom they deal.[11] Specifically, Scott argues, in order to govern their subjects, states need them to be "legible."[12] Governed subjects need to be made to fit into forms that organizations are capable of reading, processing, and relaying. This effort of fitting subjects into legible form almost always involves reduction and abstraction— typical features of information wherever it is employed. As Scott summarizes, "In each case, officials took exceptionally complex, illegible, and local social practices, such as land tenure customs or naming customs, and created a standard grid whereby it could be centrally recorded and monitored."[13] The "social circumference of official patronyms" is thus for Scott just one case of a more general project of "synoptic administrative legibility."[14]

Scott's history is undertaken with an eye to the recurring failure of such projects: his subtitle is *How Certain Schemes to Improve the Human Condition Have Failed*. For example, he discusses a project of Native American naming standardization that was announced in 1890, initiated in earnest in 1902, and ended in failure by 1913.[15] What I find of equal interest to such failures is how other such projects would in the years immediately following come to make standardized names almost obligatory. If a government project expressly charged with the task could not secure name standardization in the early 1900s, then a panoply of agencies from the 1910s to the 1930s would effectively induce such standardization as a side effect of making it the condition of participation in all manner of programs and projects. Despite beta-version failures at the turn of the century (and earlier), projects envisioning the redefinition of persons as information soon after became wildly successful.

One way to approach the critical theory of data is to show how halcyon visions of informatics are bound to fail, bound to call for resistance, bound to produce misinformation in overwhelming us with their reliable data. But I do not focus my critical genealogy on such sites, for I find more unnerving how projects of datafication bound for eventual failure can nonetheless enjoy extremely long runs of success in the interim—and despite everyone knowing their imprecision, their objectification, and their abstraction. A person is not just a name, we insist. "I am not a number," we chant. Such clichés are perfectly fitted to the informational persons we are. The real problem is that, despite all our familiar objections, we all hold dear to our names and numbers, including the numbers that we are.

We are not only our numbers, of course, but we are our numbers too. Converting a person into a number is certainly an abstraction. Yet who could deny that so many such conversions have come to define the terms of who we can be? Who would refuse to give their Social Security number to the bureaucrat at the government office or the banker evaluating a mortgage application? Who, aside from the extraordinarily wealthy who have become corporations unto themselves and the downtrodden paperless who are forced to have no choice in the matter, could even contemplate such a refusal? However, my point is not that resistance is futile or bound to fail.[16] I only seek to highlight the rarity of resistance so as to raise its stakes. The politics of data is buried so deeply within us that we tend to not notice its work. I thus aim to trace the emergence of information technologies around which quietly accumulated an entire politics of data that we find ourselves entrenched in and yet largely unwilling to attend to. This means attending, at least at the beginning, not to the genealogy of practices of resistance prophesying the inevitable crash of the data regimes we inhabit, but rather the genealogy of how those data regimes came to be a habitat we now live within without even thinking about it.

If that habitat is ours, we have become creatures of it. This is all I mean in saying that we are our data. Following Scott's interrogations of legibility, we can see that what gets formatted by data technologies is not just data, but also subjects of data, or informational persons. My argument contributes to a growing scholarly conversation about documentary identification and registration,[17] many of whose contributors also take Scott's concept of legibility as a theoretical frame.[18] This work focuses on what the editors of a volume on documentary registration aptly summarize as "infrastructures of personhood."[19] One contributor to that volume, writing about identification in nineteenth-century Egypt, observes that "it can be clearly historically documented how it was that bureaucratic forms of identity registration found necessary and expedient to the purposes of state of the governor in Egypt gave rise to a novel legal practice and ultimately to a concept of the individual, rather than any clear prior concept of the individual and their rights having called forth a registration procedure."[20] Following Khaled Fahmy, my interest in exploring documentary identification in early twentieth-century America is not that of charting how a preexisting subjectivity (such as citizenship or individuality) called forth a regime of registration as its mirror, but more that of navigating how the deployment of information technologies of registration ushered in a new mode of subjectivation.

The genealogy of documentary identity detailed below focuses on what remains today the foundation of the American system of documentary identification: the birth certificate. An initiative to standardize a set of universalizable information technologies for birth registration began in 1903 and would not be considered completed until 1933, when every state was registering 90 percent or more of its births. After tracing the contours of this history of birth registration, I briefly take up what is the obvious low-hanging fruit for any book focused on the history of informational identity—namely, registration numbers, which in the American case will mean Social Security numbers. A look at Social Security enumeration in the mid-1930s reveals a moment in which practitioners of informational identity could achieve in a few short months what had previously taken decades. Only a few years after the completion of the three-decade birth-registration campaign, the Social Security Board would assign Social Security numbers to more than 90 percent of eligible American workers in just three months. Both episodes took place in that rosy dawn of the information age. That dawn portended the twentieth-century state putting itself in the position of thoroughly relying on information technology to do its work of formatting the people it would govern, guide, and come to care for.

A Genealogy of Documentary Identity: Registering Birth, ca. 1913–1935

"Who are you? What is your name?"[21] So begins a 1919 publication by the US Children's Bureau titled *An Outline for a Birth-Registration Test*. This was a moment in which questions like these were gaining prominence as the kind of colloquy one could increasingly expect in the conduct of everyday life. The Children's Bureau counted on its reader to recognize not only his or her own ability to answer these questions, but also the need to be able to do so with evidence that could satisfy a bureaucracy: "Anyone can answer these questions, but some persons may find it rather difficult to prove the truth of their answers."[22] The

Children's Bureau was seeking to stabilize a particular piece of information technology as provenance for these needed answers: "Only the person whose birth has been registered can easily establish his age and identity."[23]

A host of organizations were at the time dedicated to installing a national birth-registration system: this included other federal agencies like the US Census Bureau, private organizations like the American Child Health Association and the American Medical Association, and even large business firms like the Metropolitan Life Insurance Company. In 1919 their efforts were just beginning to come to fruition. Birth-registration boosters could now foresee success in their project of what they called "human bookkeeping" and "the bookkeeping of humanity."[24]

Birth certificates became so crucial to the bookkeeping of persons not so much because they represent our entry into the oxygenated world outside of our mothers' wombs, but moreso because they format our first points of entry into the information systems that are the atmosphere for so much of what we do in the world. Birth certificates format many of the first stable facts about us as we travel from womb to world. And yet historian Susan Pearson recently notes that "we know almost nothing about how the birth certificate became the foundational form of identification."[25] In learning how this crucial piece of informational identification was assembled, we learn much about how our very selves can be taken to be stable today. Across three decades of assembly, from 1903 to 1933, birth certificates both "shifted epistemological authority," as argued by Pearson, and also transformed modes of subject formation and the operation of power, as I argue below.[26] Central to such shifts, I show, was the development of workable information technologies—including, most important, standard registration forms, a protocol for a registration bureaucracy, and an audit of registration success.

PRECIPITATING BIRTH REGISTRATION, 1903–1915

Early-century campaigns for birth registration began in 1903. This was the year in which the Census Bureau, which only the year before was converted from a decennial pop-up shop to a permanent government agency, published a quartet of pamphlets urging states to adopt two technologies of birth registration.[27] These were a new standard birth certificate form and a set of protocols for standardizing registration processes. The guiding details for these forms and protocols were disseminated via the American Public Health Association's model law for birth and death registration, which persists in revised form today as nationwide guidance for a multitude of vital events registration practices.

The 1903 Census pamphlets offered three distinct justifications for the adoption of the technical specifications of the model law: the utility of registration documents as legal records, the utility of registration data for compiling what were then frequently called "sanitary" (i.e., public health) statistics, and values of those data for demographic analyses of population movement. Census's pamphlet no. 100, a compendium of policy papers, research reports, and Congressional resolutions, laid the groundwork.[28] Among the materials collated in the pamphlet was a report from the American Public Health Association coauthored by Cressy Wilbur, the vital statistics chief of Michigan who would soon after assume the position of chief statistician for the Census Bureau's Vital Statistics Division from 1906 until 1914.[29] The report clearly stated the three justificatory strategies used to boost registration. It cited first "the protection of certain rights and privileges of individuals and of families" as afforded by legal registration records; it then adduced as a "subsidiary" justification "the compilation of sanitary statistics" as used by "public health services"; and third, it spoke to the importance of "knowledge of the movement of population."[30] Five years later, a 1908 Census Bureau publication presented under the heading of "Most Important Uses of Registration of Births and Deaths" the same three "reasons demanding the registration of births and deaths, stated in increasing order of importance: (1) Knowledge of the movement of population (demographic uses); (2) protection of the lives and health of the people (sanitary uses); and (3) protection of the rights of the individual and of the community (legal uses)."[31] Each of these reasons can be conceptualized in terms of how birth registration makes persons, to employ Scott's concept, legible.[32] But legibility here was not only a state project; legibility's justifications were accumulated by multiple state and nonstate agencies.[33]

Notable is that the reasons being adumbrated for birth registration beginning in 1903 were very much material of the preceding century. In 1851 the Senate and House of Representatives of the Commonwealth of Pennsylvania passed an Act to Provide for a Registration of Marriages, Births, and Deaths: "WHEREAS, From the death of witnesses, and from other causes, it has often been found difficult to prove the marriage, birth or death of persons, whereby the rights of many have been sacrificed, and great wrongs have been done: *And whereas*, Important truths, deeply affecting the physical welfare of mankind, are to be drawn from the number of marriages, births, or deaths that during a term of years may be contracted or may occur within the limits of an extensive Commonwealth."[34] Herein is anticipated all three legal, sanitary, and demographic justifications for birth registration that the Census Bureau would collate five decades later. As to registration itself beyond the arguments for it, a few states had laws like Pennsylvania's on the books even well before the middle of the nineteenth century. Yet their success was quite limited. Connecticut established a registration law in the mid-seventeenth century (as did Massachusetts and Virginia) and began requiring annual reporting in 1854—but an 1878 survey showed that "many towns made no returns whatever."[35] It was not until a 1917 revision to the Connecticut code that registration authority was centralized, mandates were imposed on professionals supervising births, and noncompliance penalties could be enforced.[36]

Though largely unsuccessful, the positivity of these laws indicates that already by the mid-nineteenth century a set of rational conditions for birth registration had been established. In subsequent decades there were repeated references to these same three rationales and occasional invocations of other kinds of justifications (such as claims for economic or business benefits).[37] Yet with all this argumentative structure in place for decades, birth registration still failed to take off as late as the concerted efforts of 1903. When it did finally consolidate, during a two-decade period beginning around 1913, it was not due to any explicit shift in rational argumentation. We do not find in these years new kinds of reasons offered on behalf of registration nor do we find a repositioning of the three kinds of reasons that had been familiar for at least a few decades. Something else, beyond rationality, was at work.[38]

One way to track such shifts as they occur below the level of official discourse is by way of the analytics of power developed by Michel Foucault. The Census Bureau pamphlets and subsequent publications offer ample evidence of a set of political rationalities matching what Foucault's work refers to as sovereign power and biopower.[39] The legal records argument for birth registration can be read in terms of the formal rights and legal command structure of sovereign power. The statistical arguments about public health and demographic analysis can similarly be read as referencing the mechanics of biopower, a politics devoted to the regulation of populations in the interests of social health. In the light of these two political rationalities, birth registration can be seen as an instrument for predefined biopolitical and sovereign purposes, and indeed as a privileged switch point or exchanger element between sovereign power and biopower.[40]

This instrumentalization of birth registration served as a kind of blackboxing of registration technologies in virtue of which the politics of these technologies were obscured at the same moment that they came to be widely used.[41] If we thus look beyond the level of explicit argumentation offered for registration, we can discern in the actual function of its technologies the emerging operation of yet another layer of power: infopower. This exercise of power implemented an implicit rationality that was overshadowed by the explicit legal-sovereign and biopolitical-statistical rationalities invoked in government publications. Yet we can excavate that implicit rationality and its correlative exercises of power out of the depths of the technologies of formatting central to emergent birth-registration practices.

The primary fact signaling the need for such a critical excavation is this: birth registration at the turn of the century still lacked a technical apparatus for implementing the official rationales widely embraced at that point. The success of birth registration in America depended not only on its official trio of justifications, but it also came to depend on a trio of information technologies. And it is these technologies that above all bear the stamp of infopower.

Two of these technologies were already under design by 1903: the standard birth certificate blank and a protocol for registration administration. Seen through the lens of contemporary technologies, we can view these as a file standard and a filing protocol. Together they enact the work of formatting in the sense I give that term here—a technique that serves to fasten that which passes through it. Standardized blank certificates and registration bureaucracies promised to fasten the babies that would be formatted according to their terms. To fulfill that promise, a third technology would need to be implemented alongside the first two: an information audit. Before turning to that crucial technology for the stabilization of informational formats and their informational persons, I first describe the way in which blank forms and handling protocols promised to fasten.

These paired technologies of blanks and bureaucracies formed a single act of formatting insofar as they were functional correlates of one another. Standard documents need standard protocols for their collection, storage, and retrieval just as standard filing protocols require standard files with which to implement their operations. Consider a purportedly standard form that nobody knew how to file. Such a document may have all the trappings of standardization, but it would not truly be standard for the reason that it could not function as such. Such a document could only be tossed on a pile with other miscellanea.

With respect to standard forms for birth registration, Wilbur retrospectively reported in 1916 that as late as the 1900 census, "no two states and but few cities in the country employed precisely the same forms of birth and death certificates."[42] This fact may seem innocuous, but from the perspectives of statistical demography, public health statistics, and legal records it was ruinous. What problems faced attempts to use birth certificates from one state in a court in another if the two recorded different data? How could data be collated if different jurisdictions not only asked different questions but, more problematic, asked similar questions in different ways? Competing definitions of "stillbirth" in the period evidence the difficulties involved.[43] Computing infant mortality rates across jurisdictions that differently define such categories would be an exercise in futility, fudging, or most likely both.

In the long campaign for birth registration, much attention was thus given to forms. This attention was expressive of what I am describing as the power of formatting. Consider, for instance, debate over seemingly innocent details of form design. In 1915, the state registrar of Virginia published an article in the *American Journal of Public Health* on "A Standard Certificate of Birth." W. A. Plecker's opening discussion may appear trivial in its concern with the durability and lightness of the paper on which the blank forms were printed, or its notes about requisite shelf space needed for differently sized forms. But someone must make these decisions and their being decided in uniform fashion is precisely what formatting is. Seven years later, the state registrar of Kansas contributed to the same journal an article titled "Simplicity in Preparation of Blanks and Forms." Its author, Charles Lerrigo, recognized that "blanks and forms are the confining channels through which flows the stream of vital statistics. Lacking their controlling direction, a surging mass of data would be hurled upon us."[44] Lerrigo presciently noted how standard forms can impose their controls on those they help form: "We must bear in mind when we decide upon uniform blanks, that we are not a uniform people."[45] Lerrigo offers an example of the different kinds of answers given to a form question asking if the registered baby was born alive and remains living: "Yes," "4 A.M," "Fine baby."[46] Another example is the use of standard forms to establish legal names of newborns: "Names are of all lengths. They must often be written by those who have no clerkly skill. . . . Let him try to crowd his letters into a limited space and he gives you little but a confused blur of cryptographic symbols."[47]

Comparing these two articles on form design, Plecker's is starkly different from Lerrigo's on a point that reminds us of the stakes of such seemingly trivial matters. Both Plecker and Lerrigo conceded the need for design changes to the form's questions to meet the needs of different populations of users, but Plecker strikingly justified this in terms of "securing complete and correct information from our eight thousand ignorant midwives, most of them negroes."[48] Devoted to subjects of technological design that so often want to be presented as neutral, Plecker's article is evidence of the racial and class-professional dimensions of technological crafting in the era of Jim Crow.

Even without this kind of politicized racialization of technology's end users, such techniques enact power. For these forms served to format and fasten all their babies. The forms produced the babies, from birth, as tied to specific data points around which their subsequent life could be accumulated, and at an accelerated rate. Some of these data points, such as ascribed race, facilitated the enrollment of their helpless subjects in enduring oppressions.

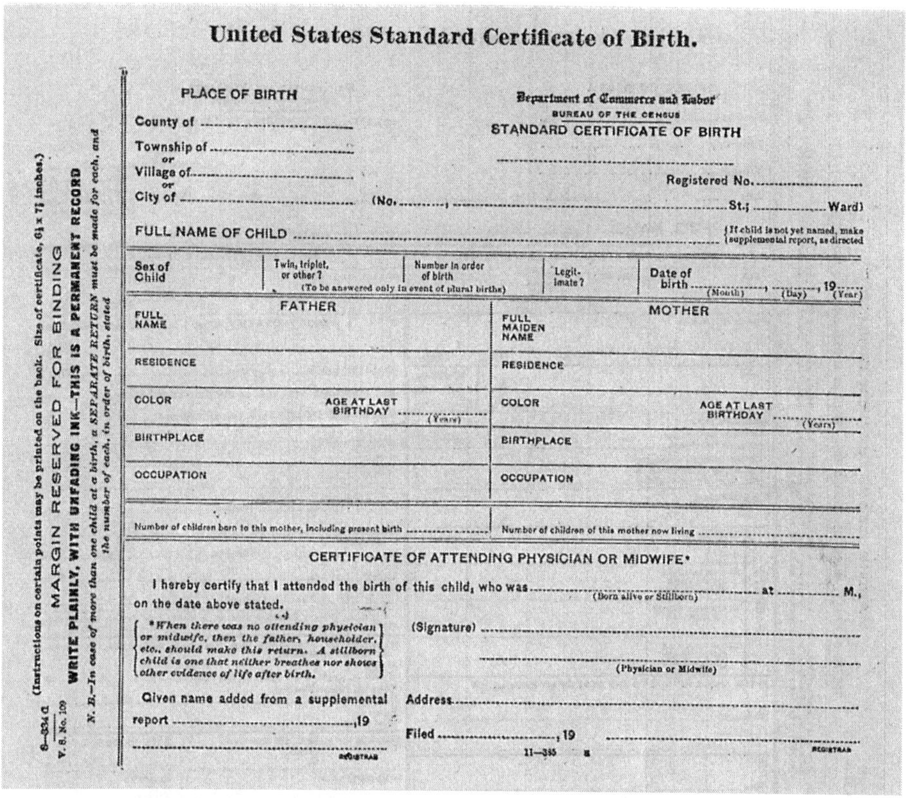

United States Standard Certificate of Birth.

PLACE OF BIRTH
County of ....
Township of ....
or
Village of ....
or
City of .... (No. ...., ... St.; .... Ward)

Department of Commerce and Labor
BUREAU OF THE CENSUS
STANDARD CERTIFICATE OF BIRTH
....
Registered No. ....

FULL NAME OF CHILD .... {If child is not yet named, make supplemental report, as directed

Sex of Child | Twin, triplet, or other? | Number in order of birth (To be answered only in event of plural births) | Legitimate? | Date of birth .... (Month) .... (Day), 19.... (Year)

FATHER
FULL NAME
RESIDENCE
COLOR
AGE AT LAST BIRTHDAY .... (Years)
BIRTHPLACE
OCCUPATION

MOTHER
FULL MAIDEN NAME
RESIDENCE
COLOR
AGE AT LAST BIRTHDAY .... (Years)
BIRTHPLACE
OCCUPATION

Number of children born to this mother, including present birth ....
Number of children of this mother now living ....

CERTIFICATE OF ATTENDING PHYSICIAN OR MIDWIFE*

I hereby certify that I attended the birth of this child, who was .... (Born alive or Stillborn) .... at .... M., on the date above stated.

*When there was no attending physician or midwife, then the father, householder, etc., should make this return. A stillborn child is one that neither breathes nor shows other evidence of life after birth.

(Signature) ....
.... (Physician or Midwife)

Given name added from a supplemental report ...., 19....
.... REGISTRAR

Address ....
Filed ...., 19.... .... REGISTRAR

8—334 d
V. S. No. 109
(Instructions on certain points may be printed on the back. Size of certificate, 6½ x 7⅞ inches.)
MARGIN RESERVED FOR BINDING
WRITE PLAINLY, WITH UNFADING INK—THIS IS A PERMANENT RECORD
N. B.—In case of more than one child at a birth, a SEPARATE RETURN must be made for each, and the number of each, in order of birth, stated
11—385

FIG. 1.1 Model Standard Certificate of Birth (US Public Health Service 1914, 93)

Other data points—for instance, length or weight—appear more innocent. But even those that appear innocent are not therefore neutral. These forms tethered their infantile subjects to the formats they exhibit—they made the persons these babies would become accessible in terms of just those formats. That precisely is the canalizing and quickening of the operation I am calling fastening.

It is therefore not accidental that there was much debate and deliberation over the specificities of the formats of birth-registration technologies. Such specificities were in many ways the entire point of these information technologies. The basic framework for debates over design was initialized in 1903 by the Census Bureau and the American Public Health Association (APHA) in their joint work on a standard form as conveyed in the same Census pamphlets described above, most notably pamphlet no. 104, *Registration of Births and Deaths*. The pamphlet includes a reproduction of a standard certificate of birth designed by the APHA Committee on Demography in coordination with Census. Though it would undergo numerous revisions in the years to come, this standard birth certificate blank remained significantly the same until the postwar period. The specifications for the form were meticulous: "Every detail called for by the standard certificate has a specific use."[49] In terms of size, the form should be "6 by 8 inches, including 1-inch margin for binding," a shape that would "[distinguish] it immediately from the certificate of death."[50]

Form size not only bears on capacities of data solicitation, such as that of names of different lengths—it also bears on filing protocols as a correlative technology of standard files. Immediately after defining their size, pamphlet no. 104 suggests that the forms can be easily filed in a cabinet with drawers 8¼ inches wide by 6½ inches deep until they are ready to be bound into volumes. When ready for binding, "five hundred of these make a small and compact volume, about 2 inches thick, and 80,000 of them, when bound, require only shelf room afforded by a space of 3 by 5 feet."[51] These seemingly trivial details are in reality anything but inconsequential. Debates today over storage capacity are conducted in the lingua franca of "gigabytes and terabytes" rather than "inches thick," and yet the square footage of server farms is just as important a feature of systems design today as was linear footage of shelving capacity for birth registration one hundred years ago.

Another filing protocol detailed in Census pamphlet no. 104 concerned the shift from a "book record" or "index" system of data storage to what we might today simply call a "form filing" system. On the older book record system (perhaps an inheritance from legacy parish registry practices), local registrars would maintain books of births and other vital events.[52] When standard birth forms first came into use, many registrars continued copying their data into books. In some cases, the original forms themselves were even discarded. From the point of view of the designers of the forms, this was inefficient and incomplete record keeping. The forms were designed to solicit exactly the amount and kind of data needed for the essential purposes of birth registration, no more and no less. To copy a subset of a form's data into a book was to create an incomplete log of data. By contrast, a registrar who copied the complete data on the form into a book would be making an "exact copy" of the file for a duplicate database.[53] This is precisely what Census and APHA recommended, albeit with the rider that such a copy should not convert the data into the different formatting requirements of a book. Registrars should simply copy the form verbatim onto an identical standard blank when copies were needed. Thus, "*no book is required*" and "to keep one is just so much unnecessary labor."[54]

Census publications detailed numerous such protocols for databasing, transmission, and reporting. At the heart of them all was a centralized administrative apparatus that pamphlet no. 104 described as "a uniform state registration service, with a central office under a state registrar, which supplies the material, formulates rules, regulations, and instructions for carrying out the law, and receives the original certificates transmitted monthly by the local registrars."[55] Every state registrar should be responsible for dividing state territory into local registration districts, appointing registrars to each district, setting a calendar for periodic reporting, and maintaining a statewide database of birth-registration certificates originated through local registrars working with physicians and midwives.

The registration protocols and standard forms just surveyed were first publicized by the Census in the above-mentioned model law for state vital statistics printed in its pamphlet no. 104.[56] It was this proposed legislation that served as the primary vehicle for the dissemination of birth-registration file formats and filing protocols. Beginning with a bill passed by the Pennsylvania legislature in 1905, the model legislation began to be adopted and adapted in state after state.[57]

Though every state eventually followed suit over the next thirty years, there was considerable delay for more than a decade. The cause of delay was already anticipated, though only implicitly, by pamphlet no. 104. The publication concluded on a somber note. The final section, titled "Checks upon the Return of Births," begins thus: "It has heretofore been very difficult to determine with any certainty the degree of completeness with which births are registered in any locality."[58] It was noted that a variety of means had been tried to check on completeness, but they all "involve[d] outside inquiry and entail[ed] considerable work" in a way that made them seem impracticable.[59] Though a few suggestions for enforcing compliance were briefly described, there was also a clear admission that techniques for a proper check upon return rates were then unavailable. This was an admission of a serious problem that could not but stymie efforts for successful registration—for in the absence of a reliable way of testing completeness of registration, nobody could truly speak of accuracy when it came to birth registration, or of success in the expansion of birth registration. All they could muster was the appearance of accuracy and success.

Perhaps this is the problem that led Wilbur, at the end of eight years of work at the helm of Vital Statistics at Census, to despair over the prospects of expanding birth registration in the United States. In a 1916 survey of the status of registration, Wilbur speculated about the dream of complete registration. He quoted himself from an 1895 address in which he had suggested that "even under the most favorable circumstances, however, we can hardly expect this result [what Wilbur referred to earlier as 'an accurate system of vital statistics

in operation'] to come about spontaneously before the middle of the twentieth century." Wilbur was quoting himself some twenty years later simply so
that he could say that he now found "little reason to revise this judgment. . . .
We can not be too sanguine in anticipating the coming of thorough and complete registration of vital statistics for the entire United States."[60] Wilbur was
dead wrong in his prediction. By 1929 all but two states (South Dakota and
Texas) had met the Census Bureau requirements for adequate birth registration, and already by 1927 most registration officials regarded the challenge
of birth-registration expansion as effectively solved. What enabled a birth-
registration project that had proceeded with frustrating slowness for an entire
decade to take off so rapidly?

In 1911, University of Pennsylvania professor Robert Emmet Chaddock sought
to summarize the present status of registration in the United States: "We live
amid a wilderness of recorded data."[61] The specific problems diagnosed by
Chaddock included the lack of a professionalized registration system,[62] the
irregularities between various state and local vital statistics reports,[63] and in-
completeness of birth registration in every single state.[64] This last problem was
the most acute. The Census claimed in 1908 that "as yet not a single state or
city in the United States has achieved successful [90-plus percent] birth registration."[65] Wilbur later interpreted the data underlying this claim as showing that "only about one-half of the births that occur each year in the United
States are recorded."[66] Similarly looking back from 1924 to the founding of the
Children's Bureau in 1912, Ethel Watters succinctly captured the "elementary
facts" of registration at that time: "We actually did not know how many babies
were born in the United States every year."[67]

Chaddock's statement of the problem is of a piece with today's talk of
information overloads and data gluts. His proposed solution is even more
interesting. Chaddock suggests that to "disentangle real information from
unrelated masses of data," to deal with the wilderness, "we need more data
and better analysis."[68] Today we are immersed in Chaddock's assumption that
the solution to any problem of untamed data must involve the development
of additional information management capacities. We hold that only data can
solve the problems of data. This, precisely, is what it means for us to live in an
"information age," and to reason within the horizon of a data episteme.

The concern with problems of data management in the early twentieth
century clearly indicates the import of the technological minutiae I have been

focusing on. The technological features of file standards and filing protocols cannot be regarded as merely accidental or innocently inconsequential elements in the design of information systems like that of birth registration. They are precisely what needed to be worked on in order to address the problem of wild, untamed data. The miscellanea of technical specs and form standards may appear unimportant in comparison to justificatory regimes invoking health and law, but it pays to remember that those justifications (and any rebuttals to them) come to naught if they are not engineered through a technical apparatus that supplies practical capacities that make the difference between an idle idea and a practice that works.

Indeed, all the promise of the birth-registration system may very well have remained unrealized were it not for a third information technology introduced beginning in 1913. Standard blank birth certificates in concert with a standardized filing protocol had given rise to growing masses of birth data. The key data management problem here was that of measurement. For birth-registration data to be useful, it would need to be shown that those data were relatively complete. What use was a public health statistic formed on the basis of an odd-shaped portion of the population? What use was a demographic figure if nobody could say with any certainty what percentage of the population it represents? What use was a legal record that only an unknown portion of Americans could even come into possession of? What was needed to negotiate such problems was a technical measure whereby reliable estimates could be made of the percentage of Americans who were registered at birth.[69]

The instrument of measure developed in response to this problem was that of the "birth-registration test," following work undertaken at the Children's Bureau beginning in 1913. Prior to this work, numerous measures had been proposed, explored, and abandoned by other agencies. In its 1908 publication on birth registration, the Census had stated the problem in exceedingly clear language.[70] The publication offered an eight-point comparison of the elements necessary for registration of deaths and births. For seven of the eight necessary elements, death and birth registration were described as identical. The one asymmetry concerned the enforcement of registration. In the case of death registration, an enforcement mechanism took the form of a legal requirement that all burial and removal permits bear the stamp of an official certifying registration of death. Without such evidence, sextons were not allowed to bury or remove their decedent. For birth registration, there was no symmetrical carrot that could function as an enforcement mechanism. The problem was acute in the first decade of the twentieth century when little functioned as a practical incentive for securing a birth certificate. Most Americans had lived their entire lives up to that point without being called on to provide a birth certificate. Most reasonably assumed that their children would so live too.

Facing such problems, a variety of enforcement mechanisms were explored: penalties for delinquent physicians, penalties for registrars whose registration districts were found to contain incomplete data, and various kinds of solicitation for registration from new mothers. A different kind of nonpunitive enforcement mechanism described by Census involved sending canvassers house to house to discover unregistered births at the door in hopes of securing registration on the spot.[71] This mechanism would be very costly in terms of labor.[72]

As these and other enforcement mechanisms were being explored, there began to emerge a number of incentives for registration such as laws, bureaucratic requirements, and commercial practices that found efficient uses for birth certificates where they existed. One of these was child labor legislation, which instantly created incentives for children of legal age to produce evidence of their eligibility to work.[73] Such laws also incidentally created incentives for children (and parents) to falsify such evidence, one of numerous kinds of documentary forgery and identity falsification then coming into being.[74]

As such incentives and mechanisms came into being, the problem of registration completeness and its accurate measurement would be felt ever more acutely. The Children's Bureau, founded as a federal agency in 1912, took the first steps toward developing a satisfactory response. Their solution involved repurposing the above-noted house-to-house canvassing method. The method, however, would not be used for actual registration. It would become a method for measuring registration. The Children's Bureau initially coordinated their efforts with Census, whose Vital Statistics Division was at that time still under Wilbur's directorship. But the brunt of this enormous labor task eventually fell to the Children's Bureau, which devised an ingenious solution to an administrative problem that other agencies had been unable to surmount.[75]

According to an institutional self-history published by the Children's Bureau at its centenary in 2012, "Beginning in 1914, Bureau staff worked closely with volunteer committees to investigate birth registration in small areas throughout the country. These volunteers were typically women, many of them members of the General Federation of Women's Clubs. . . . Their efforts resulted in the establishment of a 'birth registration area' of 10 States and the District of Columbia by 1915."[76] This narrative conveys the course of events, but its details are not quite correct: the Census Bureau's Birth Registration Area was actually established in 1916,[77] and the Children's Bureau project actually began in 1913 with work that year including planning, securing field volunteers, and initial canvassing officially reported by the bureau's first chief, Julia Lathrop, as "progressing satisfactorily."[78]

Over the next few years, the Children's Bureau project progressed.[79] What were the mechanics of the process that finally yielded a data management strategy others had failed to develop? How, in other words, was the birth-registration test conducted? Lathrop's *Third Annual Report* conveyed the details:

> The method of the test is as follows: Copies of the standard birth-registration blank are furnished by the Bureau of the Census and the correspondence is conducted by the Children's Bureau. Members of the committees receive copies of the standard birth-certificate blank, and after having carefully filled them out for a certain number of babies in their neighborhoods they then compare these records with those in the local registrar's office so as to discover in each instance whether the births have been registered and whether the record is properly filled out. The certificates are then sent to the Children's Bureau for tabulation.[80]

A house-to-house canvass in limited neighborhoods was employed to fill out blanks "for a certain number of babies" in each area. These were then taken as a kind of default sample for a broader area. A house-to-house survey of born babies moved registration measures toward a reliable baseline against which the accuracy of haphazard registration could be checked.

Further details on the mechanics of the test were provided in two subsequent bureau publications: a 1916 bulletin titled *Birth-Registration Test*, and a 1919 monograph titled *An Outline for a Birth-Registration Test* and published two years subsequent to the Children's Bureau having ceded the entire project to the Census Bureau.[81] The latter was an exquisitely detailed seven-page outline of every step involved in the test, complete with sample forms. It conveyed such instructional minutiae as how to calculate percentages, guidance on filing canvassing cards in file boxes, and the recommendation that the work of comparing the canvass data with the official data on actual birth registration "should be intrusted to the supervision of a person experienced in the use of files or in office work."[82]

What was the purpose of the test from the perspective of the Children's Bureau? The 1916 bulletin was quite clear that the tests were measures only. Though supplied with standard blanks, the volunteer canvassers were not conducting an "actual registration of a baby's birth" and as such the bulletin advised them that they should simply ignore some questions on the standard certificate (especially those that could make interviewees uncomfortable, such as questions about the "legitimacy" of the birth).[83] The 1919 monograph

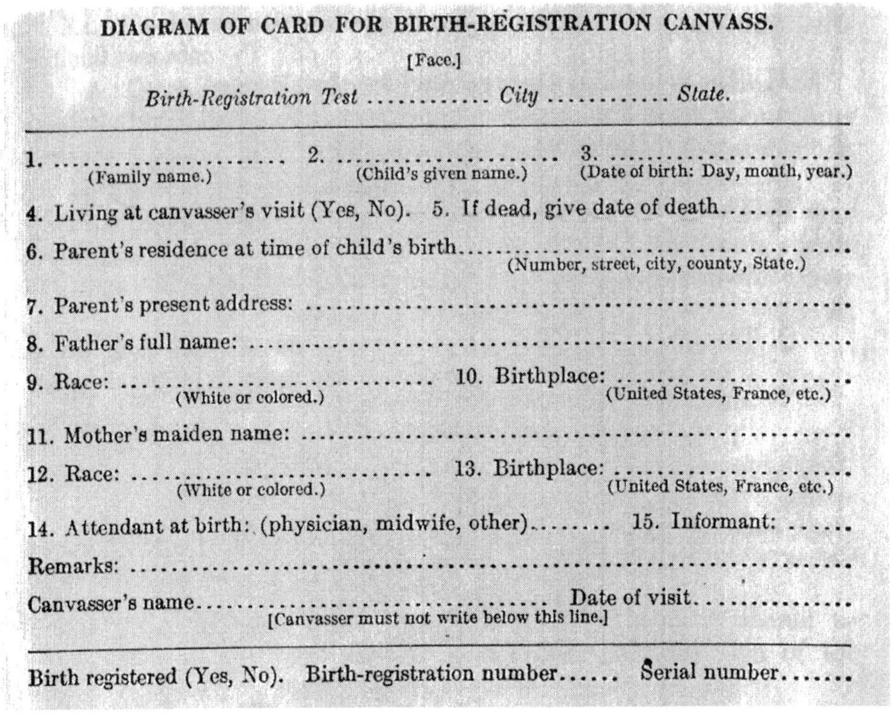

DIAGRAM OF CARD FOR BIRTH-REGISTRATION CANVASS.

[Face.]

Birth-Registration Test .... City .... *State.*

1. .... 2. .... 3. ....
   (Family name.)   (Child's given name.)   (Date of birth: Day, month, year.)

4. Living at canvasser's visit (Yes, No). 5. If dead, give date of death............

6. Parent's residence at time of child's birth......................................
                                          (Number, street, city, county, State.)

7. Parent's present address: ....

8. Father's full name: ....

9. Race: .... 10. Birthplace: ....
          (White or colored.)   (United States, France, etc.)

11. Mother's maiden name: ....

12. Race: .... 13. Birthplace: ....
           (White or colored.)   (United States, France, etc.)

14. Attendant at birth: (physician, midwife, other)........ 15. Informant: ......

Remarks: ....

Canvasser's name.. Date of visit..............
                        [Canvasser must not write below this line.]

Birth registered (Yes, No). Birth-registration number...... Serial number.......

FIG. 1.2 Standard canvassing card for use in birth-registration tests (US Children's Bureau 1919, 7)

indicated the replacement of the standard birth certificate blanks with a standardized canvassing card, thereby making it quite impossible for the registration test to function as an actual registration.

The Children's Bureau project was not a registration campaign, but an information audit.[84] As described in their 1919 publication: "A birth-registration canvass is a necessary part of a community's auditing."[85] The test was an audit of the quality of the particular formats initiated by registration forms and filing protocols. Through those formats—and only through them—the community itself could be subject to audit. This is why it is crucial that the goal was not the production of actual certificates of birth for real babies. The goal was rather to measure the success of the information systems on which the technology of birth registration had been built. Without such an audit, nobody could know what proportion of babies in a specific locale or year were registered. Without such knowledge, it would be exceedingly difficult to adjust the process of birth registration to achieve greater registration.

In developing a birth-registration audit of underlying technologies of birth registration, the Children's Bureau was working squarely within its legal mandate to "investigate and report" on all matters of child welfare and child life, and to "publish the results of these investigations."[86] This was essentially a mandate to provide and audit information. In view of their agenda, the Children's Bureau work on birth registration was in both conception and practice a baby information project.[87] Such a label may feel like an anachronism, but it is one in only a limited sense. Though Lathrop to my knowledge never described the first projects of the Children's Bureau in exactly those terms, a self-history published in 1937 at the bureau's twenty-fifth anniversary already recognized the agency as "a clearing house of information" that "assembles data on the various phases of child health and child welfare" for purposes of analysis, interpretation, and public distribution.[88]

The Children's Bureau's decisive contributions to the informatics of birth registration ran from late 1913 to sometime in 1917. At that point, the registration test project was appropriated by the Census Bureau. One historian writes that "the new director of the Census Bureau [Sam L. Rogers, who replaced then chief of Vital Statistics Cressy Wilbur with new chief William H. Davis] decided to take the lead, in the name of professionalism."[89] Another historian notes that Rogers referred to the Children's Bureau's workers as "amateurs."[90] There can be little doubt that gender was ingredient in the professionalism that the Census Bureau assumed at the expense of the Children's Bureau.[91] This was a familiar tactic being deployed by numerous agencies against the Children's Bureau.[92] The work of registration undertaken by the Children's Bureau was not only performed by a force of volunteer women, but it was also directed from within the bureau by a woman (Etta R. Goodwin) and was located in a bureau whose employees were overwhelmingly female and whose head was the first female director of a federal government agency (Julia Lathrop). Lathrop had come to the Children's Bureau after a twenty-two-year career in social work at Hull House, the Chicago settlement house run by a group of social reformers centered around Jane Addams.[93] When she came to Washington, DC, to head the new agency, one of Lathrop's first projects was the birth-registration test, an effort for which her first-wave-feminist social research work at Hull House would have provided ample training.[94] The project, expressive of what one commentator has called "Lathrop's conjoining of metrics and justice," found fast success.[95] Its being pulled from her leadership went unremarked by Lathrop in her official bureau publications. Her *Fifth Annual Report* (from 1917) was silent about the birth-registration test where her *Fourth Annual Report* had given no indication of an impending change in leadership.[96] The Census was silent on the matter too. Its first annual birth statistics report of 1917 made no mention of the Children's Bureau's contributions to the establishment of the Birth Registration Area; its second annual report failed to even mention the Children's Bureau in the context of explicit discussion of its own tests of registration completeness.[97] Following these leads, it soon became customary to refer to the Birth Registration Area as a measure established by the Census Bureau without any mention of the efforts of Lathrop's agency.[98] This silenced conflict over data management leadership indicates that not only was there a politics of the information itself, but also a much more familiar politics of information control.

In considering such complicated histories, it is important to recount both kinds of political conflict. In so doing, it is important that neither kind of politics occlude the other. This means that we need be careful to not let the kinds of politics familiar from contemporary scholarship (the gendered politics of control over information) occlude those kinds of politics for which we are still wanting an adequate analysis (the politics of information itself). This is important just insofar as the two kinds of politics in many cases depend on each other and also demand their severability from the other.

Piggybacking on the silenced success of the Children's Bureau, the Census Bureau established its Birth Registration Area in early 1916 and applied it to existing data from 1915, and then annually thereafter to each new batch of measurements. This device was used as a measure of registration at the state level. States achieving a registration of 90 percent of their births were in the area. Upon its inauguration, the Birth Registration Area included ten states—the six New England states, plus New York, Pennsylvania, Michigan, Minnesota—and Washington, DC. Following its introduction, growth in birth registration was quick. By 1927 it was estimated that 87.3 percent of the national population lived within the registration area; the following year the estimate was 94.4 percent; and by 1933 all forty-eight states plus the territory of Hawaii had been admitted to the Birth Registration Area.[99]

The Census Bureau further used the Birth Registration Area as the basis for a series of annual birth statistics publications.[100] Beginning with the fourth annual report, each volume in the series presented a cartogram of the growth of the Birth Registration Area. The Birth Statistics series for the most part published long tables of data from states in the Birth Registration Area. It presented data on infant mortality, birth legitimacy, urban versus rural births, and purported differences between various races and nationalities with respect to birthrate and infant mortality.

Alongside the Birth Statistics series there appeared similar efforts from other organizations. One of the most striking is a series of annual publications from the American Child Health Association (ACHA) running from 1919 to 1934. The ACHA's Statistical Report of Infant Mortality series presented tabular data of infant mortality based on a combination of preliminary returns from Census and its own efforts in securing data from cities not in

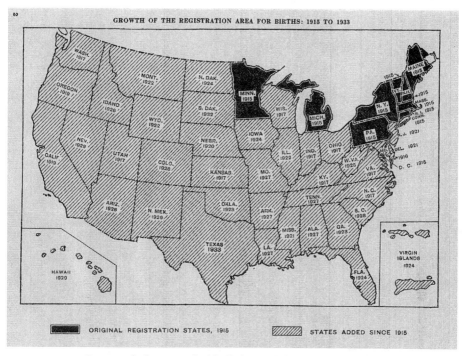

FIG. 1.3 1933 Cartogram displaying growth of the Birth Registration Area (US Census Bureau 1933, 2)

the registration area. Since it was published earlier each year than the Census series, it offered a preview of data to be later certified by Census. Additionally, where Census publications focused on state-level registration, the ACHA focused on registration at the municipal level, charting growth in the registration area from 519 cities in 1920 to 985 cities in 1933.

Included in the ACHA publication was a foldout poster-size chart displaying all cities in the registration area, with visual bars comparing their relative infant mortality rates. The primary purpose of the posters was to encourage registration in cities not yet in the Birth Registration Area. Each year's poster was flanked by text claiming that "the quickest and easiest way to reduce a high infant mortality rate recorded against your city is to demand registration of every birth, as required by law."[101] The charts are a visually rich but pragmatically unusable data device. For every year's poster that I was able to locate (1920–1922 and 1925–1929), the bars measuring infant mortality are arranged by city population, rather than by mortality rate (which would make it easy to spot the most "sanitary" and "unsanitary" cities) or alphabetically by city name or state (which would make it easy to find any given town). Only a viewer who already knew approximately where their city stood in terms of comparative population

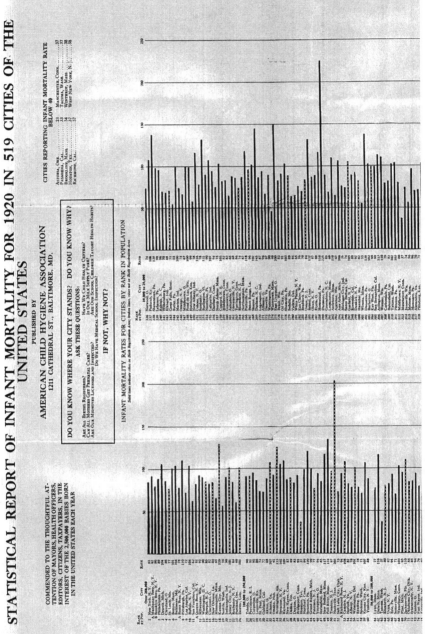

FIG. 1.4 Birth-registration campaign poster (portion), from an insert included with the American Child Health Association's (1920) *Statistical Report of Infant Mortality*

could easily find how their community was faring. The task would of course be easy for New Yorkers (the largest US city in 1920) but painfully slow for a resident of Ann Arbor, Michigan (the 294th largest city listed on that year's poster).

Alongside such informational measures deployed by Census and ACHA there emerged a multiorganization effort in boosting birth registration. The Census Bureau launched its "Every State in the Registration Area before 1930" campaign in 1924 under the leadership of Louis Dublin and in conjunction with other organizations including the ACHA and the APHA.[102] The ACHA soon became the campaign's champion, in part because of, in the words of one recent historian, "its considerable financial resources, its team of experts in social research, and an exceptional system of publicity."[103] Another dimension of the ACHA's massive influence was a function of the man who served as the organization's president from 1923 until 1928 when he left to become president of the United States. During his tenure at the ACHA, Herbert Hoover was influential as secretary of commerce under Presidents Harding and Coolidge.[104]

Put to use by a variety of organizations for what were undoubtedly a variety of purposes, the Birth Registration Area was a crucial device for stabilizing the formation of an informatics of documentary identity. What the audit process contributed to standard forms and filing protocols was a reliable measure for the comprehensiveness and effectiveness of formats. Prior to auditing, all those data collected on the forms and stored in the files could only be a wilderness that would fail to meet the demands of a field of inquiry that, in the words of the ACHA's director of research, "bristles with administrative problems that cry out for measurement."[105] With a reliable measure of completeness in hand, those data could be tamed by more data, or, more specifically, metadata in the form of data metrics.

The crucial role played by measurement in the emergence of universal birth registration is telling for the dynamics of data today. If one of our greatest contemporary promises is that of "big data," then surely one of our greatest contemporary challenges is that of big-data audits. A data set whose very size is a computational challenge must also be a set whose auditing is an even greater computational challenge. Big-data analytics promise much, but we hear precious little about how big data is being audited. Meanwhile we are confronted with set after set of data whose accuracy remains inscrutable and which is nevertheless put forward as exactly what it is: data.

"AN ACCOMPLISHED FACT," CA. 1935

The birth-registration system that still formats us today had been assembled by the mid-1930s. This is not to ignore subsequent alterations in its mechanics (including the digitization of the birth certificate system over the past few decades). Nor is it to ignore the occasional stutters and stammers that followed stabilization in the mid-1930s.[106] Nonetheless, the mid-1930s mark multiple points of consolidation: the completion of the Birth Registration Area, the dissolution of organizations like the ACHA, and above all the growing use of birth certificates as an evidential technology of identity.

The mid-1930s also saw the institutional restructuring of that wing of the federal agency that had led the long cause for birth-registration standardization. In a 1935 note in the *Journal of the American Statistical Association*, Census Bureau director William Lane Austin wrote to "Those Professionally Interested" about the reorganization of Census's Division of Vital Statistics.[107] The cause of the reorganization, explains Austin, is the division having achieved its former purpose: "For approximately one-third of a century the fundamental task of the Bureau in this field was to extend the registration areas for births and deaths. With the completion of both areas by the admission of Texas in 1933, this primary responsibility was ended."[108] Quoting an advisory committee report from earlier that year, Austin proposed that "new and intensive efforts can be devoted now to analytical treatment of the data and to the presentation of more refined results."[109] In the words of Halbert Dunn, the newly appointed chief statistician of Vital Statistics, a successful system of birth registration was now "an accomplished fact."[110]

Every information technology system is composed of at least three connected processes: an information collection and storage component, an information processing and transformation component, and an information retrieval and output component. No system can survive on the basis of any one component on its own; all three are always necessary. Of interest, then, is the relative emphasis on one or another in a given project. What Austin and Dunn clearly signaled in 1935 was that birth registration had achieved sufficient stability with respect to its input components such that focus could now shift toward a more sophisticated processing component. The success of the birth-registration project up to then had rested on ingenuity about collection and storage technologies—solicitous files, standardized filing systems, and audits for accuracy. These collection technologies managed to disprove Cressy Wilbur's early forecast that a successful birth-registration system would not be achieved until at least the middle of the twentieth century.[111] That said, Wilbur was not inaccurate with all his forecasts. Concluding his 1915 address with a discussion of the importance of birth registration, he noted the birth certificate's "increasing importance to protect the rights and insure the privileges of the individual."[112] He cited the examples of school enrollment, limits on age for child labor, and the fulfillment of pensions for widows with children. He even speculated that "perhaps old age pensions are coming."[113] Come they would—in fact, in the very year that Austin and Dunn reorganized the Census's Vital Statistics Division.

LEVERAGING POLITICAL TECHNOLOGY, 1936–1937

Congress passed the Social Security Act in 1935 with affirmative votes in the House on April 19 and the Senate on June 19. It was signed into law by its leading champion, President Franklin Roosevelt, on August 14. A suite of ten economic security programs, all but one were joint federal-state ventures, including Unemployment Insurance, Aid to Dependent Children, and Old-Age Assistance. One program stood out as different in its being designed as entirely federal. This was the Old-Age Insurance program (OAI). It is OAI that is commonly at issue today whenever Social Security is an item of political controversy.

In the early months of 1936, the newly formed Social Security Board began the work of engineering the information system that the act required it to administer. The board was given an almost blank slate. Indeed, all that the act required in terms of "administration" was that it be delegated to the Social Security Board.[114] The presidential committee report on which the legislation was based contained only a brief two-paragraph section on the topic of "Administration" in which they demurred, "The detailed working out of such coordination does not fall within the scope of this committee."[115] It thus fell to agency technicians to design the specific information technologies through which Social Security would be administered to the tens of millions of Americans who were suddenly but quietly on the verge of being fastened by these technologies.

Decisions about information technologies were just one domain of a raft of technical, administrative, and managerial challenges the new agency would face. Arthur Schlesinger, Jr., writing in the 1950s, described this work as "an administrative challenge of staggering complexity."[116] Two early chroniclers of OAI had put it this way in 1937: "This task was of a magnitude never before equalled in any Government or private undertaking, even including the United States Census, the World War draft, or the payment of the veterans' bonus."[117] With respect to the management of information, at least, the challenge was perhaps the biggest "big data" problem up to that point in history.[118] Whatever retrospective label we might wish to apply, the OAI program soon came to be known as "the biggest bookkeeping job in the world."[119] It required, its earliest architects already knew, a bookkeeping for a "volume of individual contribution records which . . . would involve administrative technique on a scale which is new to this country."[120]

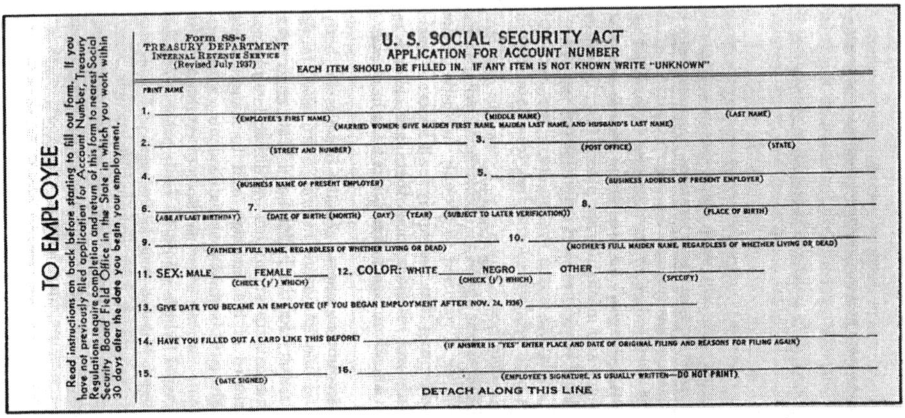

TO EMPLOYEE

Read instructions on back before starting to fill out form. If you have not previously filed application for Account Number, Treasury Regulations require completion and return of this form to nearest Social Security Board Field Office in the State in which you work within 30 days after the date you begin your employment.

Form SS-5
TREASURY DEPARTMENT
INTERNAL REVENUE SERVICE
(Revised July 1937)

U. S. SOCIAL SECURITY ACT
APPLICATION FOR ACCOUNT NUMBER
EACH ITEM SHOULD BE FILLED IN. IF ANY ITEM IS NOT KNOWN WRITE "UNKNOWN"

PRINT NAME

1. (EMPLOYEE'S FIRST NAME) (MIDDLE NAME) (LAST NAME)
(MARRIED WOMEN: GIVE MAIDEN FIRST NAME, MAIDEN LAST NAME, AND HUSBAND'S LAST NAME)

2. (STREET AND NUMBER) 3. (POST OFFICE) (STATE)

4. (BUSINESS NAME OF PRESENT EMPLOYER) 5. (BUSINESS ADDRESS OF PRESENT EMPLOYER)

6. (AGE AT LAST BIRTHDAY) 7. (DATE OF BIRTH: (MONTH) (DAY) (YEAR) (SUBJECT TO LATER VERIFICATION)) 8. (PLACE OF BIRTH)

9. (FATHER'S FULL NAME, REGARDLESS OF WHETHER LIVING OR DEAD) 10. (MOTHER'S FULL MAIDEN NAME, REGARDLESS OF WHETHER LIVING OR DEAD)

11. SEX: MALE____ FEMALE____ (CHECK (✓) WHICH) 12. COLOR: WHITE____ NEGRO____ (CHECK (✓) WHICH) OTHER____ (SPECIFY)

13. GIVE DATE YOU BECAME AN EMPLOYEE (IF YOU BEGAN EMPLOYMENT AFTER NOV. 24, 1936)

14. HAVE YOU FILLED OUT A CARD LIKE THIS BEFORE? (IF ANSWER IS "YES" ENTER PLACE AND DATE OF ORIGINAL FILING AND REASONS FOR FILING AGAIN)

15. (DATE SIGNED) 16. (EMPLOYEE'S SIGNATURE, AS USUALLY WRITTEN—DO NOT PRINT)

DETACH ALONG THIS LINE

FIG. 1.5 Social Security account number application form, Form SS-5, 1937 revision (Favinger and Wilcox 1939, 34)

Administering OAI meant designing and implementing a system for taxing employees and employers, tracking wages, and later paying out benefits to retirees (or family members). At the heart of all of this would be a technology that today appears exceedingly simple but which at the time required detailed consideration: Social Security numbers (SSNs). The first major task facing OAI administrators was thus that of registering (and assigning account numbers to) an estimated twenty-six million eligible workers. Work on registration did not begin until summer of 1936, after much of the first half of that year was squandered due to personnel issues (only fifty-three employees had been hired by July).[121] As the January 1, 1937, deadline for implementing a system for recording wage earnings approached, a system was finally adopted in the late summer of 1936.

At its heart was form SS-5 ("Application for Account Number"). Packaged with detailed instructions, this form solicited from workers sixteen pieces of personal information. The form by itself was only part of a broader solution. Using it to solicit sixteen pieces of data multiplied by an estimated twenty-six million eligible workers was a massive effort. Like the work of the volunteer force of the Children's Bureau two decades earlier, this project involved a massive information harvesting that depended on physically going door to door, and this time to nearly every residence in the country.

In a letter to the postmaster general, the Social Security Board recognized the Post Office Department as "more closely integrated with the people of the United States than any other Governmental agency."[122] The distribution of form SS-5 through letter carriers in the employ of the US Post Office began November 24, 1936. The Post Office also agreed to receive completed forms, transcribe them with its own office equipment, and send the forms and transcriptions to the Social Security Board.[123]

Just twenty-eight days after initial delivery of the SS-5, on December 22, the Post Office Department reported receipt of 22,219,617 completed applications (roughly 85 percent of the expected total). On February 15, 1937, they reported receipt of 23,647,000 (91 percent). On March 16, 1937, they registered receipt of 25,251,544 employee applications (97 percent). By August of that year, more than thirty-three million applications had been received.[124] Despite no small amount of public fear and opposition—including a front-page publicity campaign on the day before the November 1936 presidential elections—the registration project was stupendously successful.

Contrast the quick registration achievement of the Social Security Board with the work of the multiple agencies that two decades earlier had undertaken to implement birth registration. The birth certificate system involved a registration project that took almost thirty years to complete (i.e., achieve 90 percent registration nationally) from about 1903 to 1933. The Social Security registration project, initially much larger in scope, required only a few months during the winter of 1936–37 to achieve the same level of registration.

Much of the success of Social Security registration was owing to its deployment of recently emergent information machinery. In September 1936, International Business Machines outbid Burroughs Adding Machine Company and one other company to win a contract to supply mechanical equipment to the board's records office.[125] These were the machines on which were built the tabulation processes that automated record keeping for millions of workers in the form of punching holes into cards.[126] These machines occupied over an acre of floor space in the Candler Building,[127] the Social Security Board's "huge accounting factory" in Baltimore.[128] One contemporary witness considered all that record-keeping apparatus as "one of America's seven wonders of the Machine Age."[129]

Looking beyond such ephemeral wondrousness, what deserves attention are the technological processes adopted by the Social Security Board that set a course for the present. From this perspective we can recognize that OAI was one of the most successful projects of its decade to make use of the stable platforms of documentary identity that had just been elaborated over the previous two decades. This is evidenced not just by the fact that the Social Security Board endorsed birth certificates as a gold standard for proof of age.[130] The more crucial point is that the Social Security project represented one of the first constructions of a large-scale bookkeeping effort that would from its very inception be in a position to rely on a stable informatics of documentary identity. The basic assumptions of these informatics strategies would remain

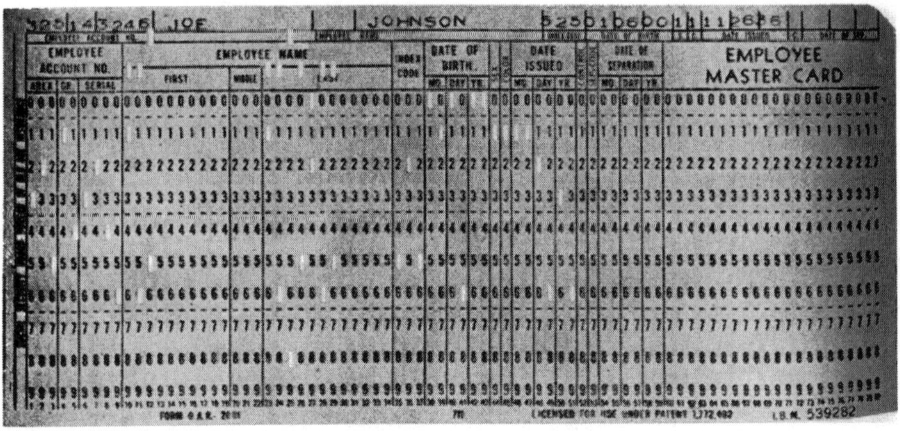

FIG. 1.6 Social Security account employee master card (Van Boskirk 1939, 22)

in place long after the original tabulating machines and punch cards had been retired in favor of newer and faster models, and those updated machines retired in turn in favor of digital computers and magnetic storage assemblies.

The early architects at the Social Security Board ushered nascent technologies of documentary identification to their practical end point, for they established a kind of documentary identification that was purely informational.[131] The SSN was an informational marker of identity that was made to operate without leaning in any way on forms of biometric certification to do its work. It thus pushed to its conclusion a possibility that the birth certificate had announced but not realized. Passports had relied on photographic resemblances, signatures on unique motions of the hand, fingerprints on the physics of our digits. Birth certificates began the process of a more minimal endocumentation of living data: parentage, date of birth, sex (the original standard forms did not even measure length and weight). The SSN inaugurated the next step of a purely informational identification: the use of identifiers that do not refer to anything other than the unique information that they are. SSNs do not represent us, index us, or resemble us. They are identifying information in its pure form of nonrepresentational arbitrary articulation.

In some sense, of course, all identification is ultimately conventional. But where we might expect that a fingerprint or a photograph bears a representational relationship to the person it is being used to identify, it would never even occur to anyone to suggest this about a nine-digit number. Even if all identification techniques are formally conventional at bottom, the actual content of most such techniques relies on a convention of representation. There can be no such conceit about the content of an account number. With the

SSN there solidified—as a correlate to the innovation of a purely informational technology of identification—a stable kind of person who could move to the beat of the drum of data.

The Social Security Board was not the first big organization to use purely conventional account numbers. Their decision to do so was based in part on the successful use of account numbers by life insurance companies.[132] But it would be a mistake to conclude on this basis that the informatics apparatus of Social Security was simply more of the same. For there was one crucial difference: Social Security needed to be implemented as a universal technology that all (eligible) workers would be registered into. No insurance company has ever dreamt of attaining every (eligible) insured as a customer. This difference may seem slight, but it is that most crucial of differences between an information practice that is content to be localized and an information practice that wants to be universal.

In the face of such requirements, and on the basis of numerous contingent decisions of design, the specific technique finally settled on was the arbitrary nine-digit numerical account number that we know well today.[133] The design specifications were the work of Elwood Way, chief of Division of Records.[134] Like many other early Social Security technicians, Way had come to the Social Security Board from the Treasury Department.[135] In addition to SSNs, he was responsible for the design of form SS-5, form OA-702 (the office record transcription of the data on SS-5 and also the form containing the Social Security cards with preprinted number that would be detached and sent to applicants once the office record was completed), and many other agency blanks and records.[136] According to Way's design, the first three digits of the SSN were an area number local to where an employee first registers (parceled out on a state-by-state basis until 1972); the next two digits were a group number useful for mechanical efficiency and verification; and the final four digits formed a serial number.

This specific formatting, applied to a huge swath of the American population in only a few short months, is too often papered over in discussions that treat matters of information design and architecture as merely technical and therefore apolitical. Beneath these formats is what might be called the technological truth of the political. I have recounted such monumentally tiny details of technical design because they can be so monumentally consequential despite being so tiny. We can continue to insist that there is nothing of political and ethical significance in these decisions of technical design. Or we can adopt an alternative view according to which technical design decisions are themselves a terrain on which politics and ethics get played out, won or lost, and then embedded in ways that canalize future possibilities. Every

device that today promises a revolution of data is at the same time the promise of the maintenance of techniques of datafication already assembled in the
past. Our contemporary data revolutions are not so much hallowed visions of
a future as they are halcyon dreams of the days when data was actually being
given the power that is now being played out by the scripts of the latest apps,
gadgets, start-ups, incubators, and collaboratories.

This lesson holds true for the forms of documentary identification I have
been exploring. It is true in the case of the birth certificate, which already by
the 1920s was becoming the "breeder document" it remains today, meaning
that it is at the root of nearly every piece of identity paperwork most Americans accumulate. It is true too in the case of the SSN. In the decades since the
first SS-5s were completed, organization after organization has leveraged the
technological apparatus of the SSN.[137] The increasing use of a number was all
that was needed to make that number a marker of identity.

Consider some of the recent consequences. In the immediate wake of a
2017 credit reporting agency data breach exposing the SSNs of more than
half of Americans, a White House cybersecurity coordinator suggested the
possibility of retiring the SSN as a personal identifier in favor of enhanced
identification technologies like cryptographic keys.[138] As government and
corporate leaders contemplate new account designs—that is, new identity
formats, leveraging new information technologies—they stand before a challenge of staggering complexity that returns us to the initiating scene of SSNs:
convincing a gigantic population of users to adopt a technology with which
they have little or no familiarity.

"The Sum Total of Your Data"

Half a century after the data of documentary identity began to become a part
of everyday life, an innocent technician in Don DeLillo's *White Noise* (historians of information theory will take careful note of the title) could casually
inform the novel's protagonist that "you are the sum total of your data." The
next line is unequivocal: "No man escapes that."[139]

The sum total of the data that we have become includes, of course, much
more than the information formatted to fit our birth certificates and Social
Security cards. But the terms by which the birth certificate and the Social Security account became stable universalizable technologies we are all expected
to use were also the terms by which that very kind of technology became
universalizable. The story of birth certificates and registration numbers in the
American context is one that therefore repeats time and time again in countless contexts—with crucial variances, of course. Those contexts form for us

a different kind of birth: our delivery into databases. Those databases are the swaddling that ever comforts the informational persons we have become.

Few, if any, of us manage to wriggle out of all that obligatory documentary identity technology for any longer than the briefest of times. Working in concert with that documentation today are numerous other aspects of identity that have been subject to datafication over the past century. The next chapter turns to the central place of data in that which seems to so many of us most ours: the mind. The mind itself has been subject to the imperatives of data, not only in the familiar forms of intelligence algorithms like the IQ or the SAT, but also in the scientific measures spawned by a domain of psychological research that subtly grasps what seems most central to a who a person is: their very personality.

Algorithmic Personality:
The Informatics of Psychological Traits, 1917–1937

Uploading Personality

Even the most committed informational luddites acknowledge that registration forms, account numbers, and other such externalia are increasingly central to how we live. But when it comes to what has long been supposed to be inside of us, the internal domiciles of the self, we frequently encounter the objection that data surely cannot tread here.

Sebastian Thrun faced this objection in 2015, when the Google X founder and tech entrepreneur floated to a Stanford symposium the idea that "perhaps we can get to the point where we can outsource our own personal experiences entirely into a computer—and possibly our own personality." Thrun deftly anticipated, even invited, his audience's shock: "It's maybe unimaginable but it's not as far off as people think. It's very doable."[1]

There is an irony in Thrun's proposal and the skeptical response it received both at the symposium and later online, for there is a rather straightforward way in which Thrun is quite right that loading personalities into computational machines is quite "doable." That it is doable is demonstrated by the fact that it has already been done. The concepts of personality we trade in today are largely artifacts of an informatics apparatus assembled one hundred years ago.

The science of personality, known as personality psychology, secured its authority from the 1910s to the 1930s by distinguishing its informatics-centered methods from the interpretive methods of other psychological approaches, most notably psychoanalysis. Personality research achieved its scientific success by virtue of its ability to stabilize new objects of study: personality, personality types, and personality traits. These objects were made stable as computable correlates of an informatics apparatus that has since grown massively more sophisticated. Key to that assembly was the work of algorithmic processing deployed on an information technology we would today regard as quite quaint:

paper-and-pencil questionnaires. Those questionnaires could produce masses of data that, after algorithmic processing, could be used as new forms of scientific evidence that stood in effective contrast to the live demonstrations, photographic depictions, and narrative representations invoked as evidence by competitor programs in psychology.

Thrun's provocation thus begs for a historical perspective. Such now-familiar personality traits as introversion and extroversion were from their very beginnings produced as the known correlates of a technical arrangement of forms, algorithms, and graphs. Excavating this history can inform contemporary anxieties about aspects of our selves we wish to think of as beyond the grasp of data. It can also spur skepticism toward the common idea that information is always a representation of something preceding it. Rather than assuming that information need be adequate to something prior to it, which it therefore will inevitably fail to reproduce, perhaps we would do better to investigate how information technologies succeed in stabilizing both the data that is their content and whatever that data is a correlate of.

This would mean accepting at the outset the veracity of psychometric technologies so widely deployed today. Although it is easy to cast doubt on the accuracy of these instruments, it is really their functional success that should impress us. Consider an example not all that far from Thrun's Google X. In the late summer of 2016, the Republican presidential campaign contracted with the firm Cambridge Analytica for its suite of data science and information operations products. Cambridge Analytica's consulting is built around a massive data warehouse that includes up to five thousand data points each on 230 million Americans.[2] Records of at least fifty million persons were harvested from Facebook by way of a personality assessment program developed by a psychology research associate at Cambridge University. The firm's data collection tools, as well as its processing engine for rendering this data useful, relied on such psychometric staples as the "Big Five" inventory that is today the consensus model for scientific personality trait analysis.[3] In excavating the histories of such psychometric inventories, I shall assume as my starting point that they are as effective as these recent deployments demonstrate. Of course, what exactly they are effective for is the crucial question. In light of their having been recruited to help pull off the most stunning American presidential election in recent history, this is precisely the question we need to begin asking.[4]

Persons with Personalities

When prompted to say who we are, most of us will offer our name and perhaps our profession or familial roles. When prodded for more, many will quickly turn toward conveying a series of self-ascribed personality traits. I am curious (or not), kind (or not), anxious (or not), or extroverted (or not). Such descriptors help constitute the core of who we take ourselves, and others, to be. They speak deeply to who we are.

Yet these features of our selves are not timeless. In 1936 psychologist Gordon Allport, the first grand technician of human personality, published a list of 17,953 personality trait terms.[5] Researchers ever since have been working hard to pare down that list into something more manageable. A dominant contemporary approach boils personality down to five core traits. Regardless of how many personality traits we have, or others take us to have, almost all of us tend to think of these traits as somehow ineluctably us. In part due to the scientific legacy of Allport's work, we now comfortably recognize ourselves in the apparatus of concise two-dimensional personality profiles on which we can be plotted or charted according to our answers to a finely tuned barrage of questions. The science of personality psychology that produces these profiles relies on a technical apparatus of questionnaires, algorithms, and graphs that make it an eminently informational science. While few of us treat our personality traits with scientific rigor, even the most casual of trait ascriptions relies on planks of informational assessment perfected by personality psychology in the first decades of the twentieth century.

Psychology emerged as a science in its own right in the late nineteenth century by carving out its independence from the speculative philosophy of which it was formerly a branch. It then rose over the next few decades to its current position of cultural authority. Psychology offers an expertise that is simultaneously epistemic and political. Psychology knows us and as such is capable of governing us—in knowing how we are likely to act, it can instruct us in how we ought to act. It is possessed of undeniable influence exemplifying Foucault's signature theme of "the interweaving of effects of power and knowledge."[6]

Writing in Foucault's wake, Nikolas Rose argues that psychology "has played a rather fundamental part in 'making up' the kinds of persons that we take ourselves to be."[7] Rose's approach is a compelling precedent for my argument in the way he approaches selfhood genealogically: "Subjectification is not to be understood by locating it in a universe of meaning or an interactional context of narratives, but in a complex of apparatuses, practices, machinations, and assemblages within which human being has been fabricated."[8] Rose's genealogy further anticipates mine in its content—its focus is the production of psychological selfhood through an arrangement of techniques centered around the crucial technology of the test: "The ritual of the test, in all its forms and varieties, has become central to our modern techniques for governing human individuality, evaluating potential recruits to the army, providing 'vocational guidance', assessing maladjusted children—indeed in all the practices where decisions are to be made by authorities about the destiny of subjects."[9] What work did the test perform in order to be invoked in all this decision-making? Rose argues in an earlier book that "the psychological individual was specifiable only to the extent that it was constituted as both measurable and differentiable."[10] Rose's claim is that the test became a central technology for the production of psychological selfhood in part because the test was a successful tool for measure. My claim in turn is that the test could produce psychological measure because it was made to wield an algorithmic apparatus that came to be taken to be effective for precisely those purposes. The test became an efficient package that could both be easily applied to *an* individual being tested and also mobilized to be quickly applied to *any* individual on the basis of the universalizability of its algorithmic computations.

Other aspects of my argument are informed by the Foucauldian investigations of Arnold Davidson, specifically his self-described "archaeological" and "genealogical" studies of late nineteenth-century notions of perversion.[11] These studies show how the concept of sexual perversion came to be seen as no longer anatomical but as functional expressions of an underlying sexual instinct.[12] Richard von Krafft-Ebing had argued in his 1886 *Psychopathia Sexualis* that gaining scientific grip on perversions of the sexual instinct required that "one must investigate the whole personality of the individual."[13] As Davidson relays, this kind of claim involved "the inauguration of whole new ways of conceptualizing ourselves" which had to do with something that soon "went under the name of *personality*."[14] What was inaugurated by the new psychologies of the late nineteenth century (be it popularly therapeutic or rigorously scientific) was the possibility of conceptualizing ourselves in terms of personalities. We began to become persons with personalities. Though the scientific psychology of personality would not consolidate until the 1920s, its earliest outlines were already visible in the 1870 to 1905 period central to Davidson's narrative.[15] Where Davidson's history recounts the earliest moments of personality, I am focused on the consolidating period when personality became a viable object of scientific measure. This field in focus enables me to situate the emergence of personality psychology in the context of a broader cultural shift, from nineteenth-century conceptions of character to a twentieth-century notion of personhood focused around an equally loose, but ultimately more scrutable because measurable, conception of personality.

Nearly one hundred years later, contemporary personality psychology defines the notion of personality in terms easily recognizable to those on our side of that cultural shift: "An individual's personality consists of *any* characteristic pattern of behavior, thought, or emotional experience that exhibits relative consistency across time and situations."[16] This is the authoritative, and easy to parse, claim of personality researcher David Funder in a textbook that enjoys wide respect in the field. Elsewhere in his textbook, Funder claims, in a line that is at once remarkable and banal, that "in some real sense people *are* their traits."[17] Funder is surely right if he means to suggest that most of us have quite simply become our personality traits—these traits increasingly define constituent parts of who we are. Nobody claims that these traits define, constrain, and determine everything about us. Yet everyone seems equally convinced that they nevertheless do tell us something important, maybe even essential, about ourselves. Even if you object to this common sense, perhaps as a matter of philosophical principle, it is undeniable that this sort of idea is in possession of enormous influence in the context of contemporary cultural conceptions of the self. As Randy Larsen and David Buss ask in their textbook, "Aren't we all curious about the characteristics people possess, including our own characteristics? Don't we all use personality characteristics in describing people? And haven't we all used personality characteristics to explain behavior, either our own or others?"[18] Every discipline is eager to find ways of generalizing its parochial concerns to the benefit of all. But in the case of personality psychology, this generalization may actually be correct. It is an empirical supposition, but I would suppose that the vast majority of Americans do self-consciously describe themselves and others using the vernacular of trait psychology. Personality traits, from their humble beginnings in 1917, have become in just one century a near-universal feature of the kind of persons we are today. We have become our personalities.

How, though, did we become persons with personalities given that at one time nobody took anybody else to have them? One way to answer this question is by way of the lead of Rose and Davidson—namely, through careful genealogical attention to the operative techniques and functioning concepts of psychological science.[19] Following their precedent, I argue that becoming our personalities involved, in part, becoming our data too.

A Genealogy of Personality Informatics:
Measuring Traits, ca. 1917–1937

Two tendencies in late nineteenth-century human sciences paved the way for the early twentieth-century science of personality: research on multiple personality in abnormal psychology and the growing use of quantitative measure in the human sciences in general. The first of these vectors was focused on the mobile specter of multiple personality as it traveled from France to America in the 1880s and 1890s when the work of French psychologists gained influence in America through the efforts of Boston psychologists like Morton Prince and Cambridge professors like William James. At the same time, a second scientific vector saw increasing use of the measure of persons, or anthropometrics. Anthropometry had for decades enjoyed prominence as a means of assessing entire populations of persons, but it was not stabilized as a science of personal measure until worked over in the 1880s and 1890s by the famed British scientist Francis Galton. To the sciences of his day, Galton contributed a craft of the measurement, recording, and analysis of persons. The first truly massive employment of such measures was then developed in the 1900s and 1910s in the work of intelligence testing. The general style of testing launched by the measure of intellect could then hook up with the study of personality inaugurated by the abnormal psychologists: at that meeting point would emerge the science of personality psychology.

The new scientists of personality proposed theoretically rich conceptions. But my interest is chiefly in their lasting influence as technicians of the measure of personality traits. In much of science, the discursive deeds of grand theory gain their force through the practical elaboration of tiny techniques. In the case of personality psychology in the 1920s and 1930s, these techniques were used to craft new kinds of persons appropriate for and amenable to psychometric evaluation. Personality researchers busily foisted a bevy of humble devices, instruments, and techniques underwriting their grander theoretical thrusts: intelligence tests in education and industry, psychological evaluations in the military, and all manner of printed forms for assembling information about persons.

PRECIPITANTS OF PERSONALITY PSYCHOMETRICS, 1883–1916

In his landmark 1890 treatise *The Principles of Psychology*, Harvard physiologist, psychologist, and philosopher William James began the long labor of introducing the idea of personality into the vernacular of American psychology. James's most thorough discussion of personality was his account of split personality as then explored by contemporary French abnormal psychologists.[20] Chief among them was Pierre Janet, whom James cites in a long discussion of "alternate personality."[21] Ian Hacking notes that personality was actually "doubled" and "multiple" before it was unified.[22] Personality was not at first the object of normal psychology capable of the kind of unity that we have in mind today when we speak of a person's personality.[23] James's own thought expressed this transition in his contribution "Person and Personality" to *Johnson's Universal Cyclopaedia*: "All these facts have brought the question of

what is the unifying principle in personality to the front again."[24] This statement marks one of the first moments of personality's transition from abnormal to normal psychology, and from multiplied or dissociative to integrated
and coherent personality. Yet decades later it would remain the case that, as
one historian of the idea writes, "as late as the 1920s the term 'personality' was
still used predominantly in discussions of abnormal psychology."[25] James had
here, as on so many others subjects, arrived too soon.

This was in part because in James's day the dominant model of selfhood
was still that of character. According to a remarkable essay by the historian
Warren Susman, a significant shift in American culture occurred at the turn
of the twentieth century when conceptions of personhood cast in terms of
character began to be increasingly replaced by ideas of personality: "Interest grew in personality, individual idiosyncrasies, personal needs and interests."[26] This displacement involved a whole series of transitions, two of which
stand out. First, there was a shift from holistic conceptions of character to a
notion of personality as a composite always in need of unifying refinement.
Historian of psychology Kurt Danziger notes that, "unlike 'character', whose
reference was to something considered essentially unitary, 'personality' was
essentially diverse, an assembly of various tendencies."[27] A second shift has
to do with personality being amenable to quantitative empirical scrutiny in a
way that character is not—at least, not as readily. This involved a shift from
a normative notion of character to a purely descriptive notion of personality.
Character was something that one either did or did not have, and it was a
good thing to have it. But everyone has personalities, and they assume different shapes in different persons such that those differences can be descrip-
tively interrogated.

Susman's history of the character-to-personality transition focused on
vernacular representations in popular psychology such as the mental hygiene
movement and other forms of what we would today call "self-help." We can
enrich this account by shifting focus from pop psychology to psychological
science. The scientific literature was at first less influential than wider popular manuals, but it eventually came to exert an influence that remains markedly massive. To achieve that influence, scientists of personality realigned the
concept of personality with emerging cultural expectations by moving it over
from abnormal psychology and into a scrutable unit of normal psychology. If
personality was split first and whole only later, Hacking notes that the "wave
of multiples had almost completely subsided by 1910," at least in France, and
then soon after in America.[28]

Among the American abnormal psychologists opening up alternative researches was Morton Prince.[29] Prince gained fame for his 1906 account of the

multiple personalities of a pseudonymous Miss Beauchamp, *The Dissociation of a Personality*. Readers were drawn in by the heroic narration of a princely doctor wrestling with "B's" many alternates including the devilish personality of "Sally," all culminating in the restoration of "the Real Miss Beauchamp" in the final chapter.[30]

What was Prince's concept of personality in his account of its dissociation? Prince's earlier work had relied on a characterological concept of feeling as capable of holding personality together.[31] Later, in his 1914 book, *The Unconscious: The Fundamentals of Human Personality Normal and Abnormal*, Prince replaced characterological feeling with a set of information-rich metaphors.[32] Personality was now "a complex affair" definable in terms of combinations of present and past states of consciousness.[33] If earlier he had argued that "different sensations compounded" accounted for this complexity,[34] the focus now was on the physiological basis for combination in what Prince called "neurograms."[35] This new word conveys the later Prince's idiosyncratic definition of the unconscious. The term, he says, was coined in reference to communication technologies such as the telegram and the phonograph then dominant in the cultural imaginary: "Though our ideas pass out of mind, are forgotten for the moment, and become dormant, their physiological records still remain, as sort of vestigial, much as the records of our spoken thoughts are recorded on the moving wax cylinder of the phonograph."[36] Prince's neurograms are mechanisms for the "conservation" (or storage) of "registered" (or formatted) experiences that would otherwise dissipate.[37] Neurogrammatic storage "complexes" form a ground for selfhood and are "a fundamental of personality."[38] On this basis, Prince advised that for therapeutic purposes it is essential to gain "exact information" of "dormant memories" that remain factors in personality.[39]

Though he did not name Freud in the book, Prince contrasted their therapeutic approaches.[40] Prince shared with Freud not only a field of research in abnormal psychology but also its attachment to the single case as the unit of therapeutic intervention and the case study as the genre best suited to conveying scientific prestige. But Freud viewed repressed memories as messages to be interpreted and thereby modified in the act of hermeneutic revelation: what matters for this theory is, estimates Prince, "bringing them to the 'full light of day.'"[41] Prince's neurograms were in contrast messages to be reformatted by an ongoing "art" of analysis. He referred to the materials for this art as "exact information of *what* we need to modify."[42] Freud saw himself as an analyst in search of meanings hidden behind information, whereas Prince saw himself as what we would eventually come to call an information analyst unconcerned with the problem of meaning.

Prince's approach would prove itself quite amenable to the rising fever for quantitative measure. Indeed, his work became a vector for later measurements of personality. Meanwhile, Freudian psychoanalysis became one of psychological science's major casualties. In his 1926 "The Question of Lay Analysis," Freud had looked forward to psychoanalysis as something more than a curiosity of the intellectuals, as something widely deployed by future "social workers" and "analytic educationalists."[43] Freud dreamt of a psychoanalyst in every school, every social insurance office, and every other major institution devoted to therapy. But psychoanalysis was never to gain such cultural gravity. Its influence is not nothing, but its modicum of pop appeal quickly pales beside the enormous institutional influence of personality psychology. How many batteries of tests of personality have your children been submitted to? What assumptions have been made about them by school psychologists because of these tests? And what of you yourself, when you were a child, or a job applicant just out of college, or one of the millions of Americans diagnosed with depression or anxiety?

Summarizing a more complex history, there were two competing lineages descending from the earliest studies of personality in the field of abnormal psychology: a hermeneutics and an informatics. Of these two, hermeneutic psychoanalysis has received enormous attention from critical theorists and cultural critics over the past century. Yet its actual influence is orders of magnitude less than that of the informatics of personality. The critics' persistent obsession with psychoanalysis may be a symptom of something deeper, but it is also a positive expression of a neglect of those domains of psychology in which something more is at stake.

The lasting influence of scientific personality research flowed in part from first-generation researchers' success in disinterring key ideas from abnormal psychology and merging them with psychometric apparatus: measurement, statistics, correlations, and other quantitative analytics packaged into off-the-shelf algorithms. Personality psychology eclipsed the hermeneutic tactics of psychoanalysis by replacing subjective narrative studies with the objective sheen of numbers, graphs, and other informational tableaux. There are few better representatives of that quantitative objectivity of human measure than Francis Galton.

Galton was a child prodigy and half cousin to Charles Darwin. Darwin gave us the central concepts and metaphors for evolutionary theory. Galton soon after helped solidify the statistical apparatus without which that theory could today hardly call itself a science. Quantifying measure was the golden thread running through Galton's diverse forays: fingerprinting, the measure of the body, the pursuit of the normal (bell) curve, and the researches in mental measure best represented in his 1883 *Inquiries into Human Faculty and Its Development.*[44] As Stephen Jay Gould observes, "Quantification was Galton's god."[45] Galton would have appreciated the compliment. His 1879 essay "Psychometric Experiments" held that "until the phenomena of any branch of knowledge have been subjected to measurement and number, it cannot assume the status and dignity of a science."[46] In a short article titled "Measurement of Character" published five years later, he went as far as recommending "measuring man in his entirety."[47]

Galton is perhaps the first technologist who both dreamt up and carried out research programs of the sort that I envision as expressive of the power of information. If infopower is normatively ambivalent by spawning both political injustices and moral improvements, then Galton is a most apposite representative. For in him was combined both the man who pioneered eugenics (he coined the word)[48] and the man who introduced to statistical inquiry the undeniably useful concept of correlation.[49] Galton's was a program of a usable and dangerous technics of, in his words, "grasp and measure."[50]

If Galton forced the view that the imprimatur of science depends on measure, then the generation of scientists following him could operate with a secure assumption of a need for measure. Psychologists were no exception. One surveyor of psychology in 1932 described the time as "the era of measurement."[51] The following year, Robert Sessions Woodworth would claim that, "in order to make a scientific study of individual differences and of their causes and effects, we need to have ways of measuring the individual's behavior and characteristics."[52] Such measurement, he said, "gives the investigator definite information to work with."[53] The decades preceding this statement saw a massive increase in the pace of production of new scientific tools of mental measure, most notably tests.[54] Bibliographies published in the 1920s surveyed these new instruments, some summarizing the new methods and others simply listing and counting them.[55]

Intelligence testing exemplifies the period's convergence of scientific measure and psychological investigation. The first robust intelligence tests were developed in France by Alfred Binet.[56] A student of abnormal psychology under Jean-Martin Charcot at La Salpêtrière in Paris, Binet eventually switched focus to developmental psychology. He first proposed intelligence testing in 1905 amid education reform movements arising out of earlier legislation mandating universal education.[57] One crucial reform problem concerned what to do with children who were not medically indisposed but nonetheless evidently unprepared for classroom work alongside other students. This was the background for a debate between Binet, the self-styled psychologist, and the opposition he met with from educational psychiatrists. The psychiatrists sought to resolve the problem by developing instruction within asylums and mental institutions. Binet sought to construct a measure whereby normal and abnormal students could be placed on a continuum in such a way as to foster support for special classes for "feeble-minded" students within the normal education system. It was a debate—one we are still having today—between testing and medicalization.

Though his success in France was mixed,[58] Binet's work became the basis for a long-term bond between psychology and pedagogy. Lewis Terman and his Stanford colleagues soon translated the Binet-Simon test into the American context to produce in 1916 the Stanford-Binet intelligence test.[59] Terman's test made safe the assumption that intelligence is a measurable piece of mind amenable to such informational operations as statistical comparison and data tabulation. This assumption contrasted to psychiatric approaches more literary and medical in perspective. As Serge Nicolas and colleagues state in a recent study of Binet, the "larger (now invisible) contribution was, simply put, to *make psychological* the question of identifying *abnormal* children in *schools* using scientific methods."[60] We might, I am suggesting, place italicized emphasis on "scientific" as well.

In the American context, emergent intelligence testing at first had an unsure place in the growing psychometric literatures. Guy Whipple's 1915 *Manual of Mental and Physical Tests*, a perfectly ordinary book in its day, placed intelligence tests at the end of its study with a cautionary prefatory remark: "The tests of this chapter differ from other mental tests described in the present volume in that they measure, not the efficiency with which certain typical mental activities or mental processes can function, but rather the number of ideas that an individual possesses."[61] Whipple's conception of intelligence testing was not quite that which would soon solidify and with which we are familiar today. Whipple thought of these tests as measuring "what he ['the individual'] knows about."[62] But intelligence tests were less a measure of amount actually known and more a measure of amount potentially knowable. They were part of a shift from the quantification of what one does to the psychometrics of who one is. Interestingly, Whipple also explicitly omitted discussion of the Binet-Simon tests, explaining that "the number of published investigations bearing upon the Binet tests is so enormous" and citing a 1914 bibliography of 254 titles on the subject.[63] Whipple had yet to really internalize the link between pedagogy and psychology that Binet, Terman, and others were forging. I point this out not as a failing on Whipple's part, but as an indicator of the uncertain status of tests of mental measure in the 1910s. Similarly, Shepherd Ivory Franz's 1912 *Handbook of Mental Examination Methods* contained the usual final chapter "General Intelligence" as well as a prefatory note announcing

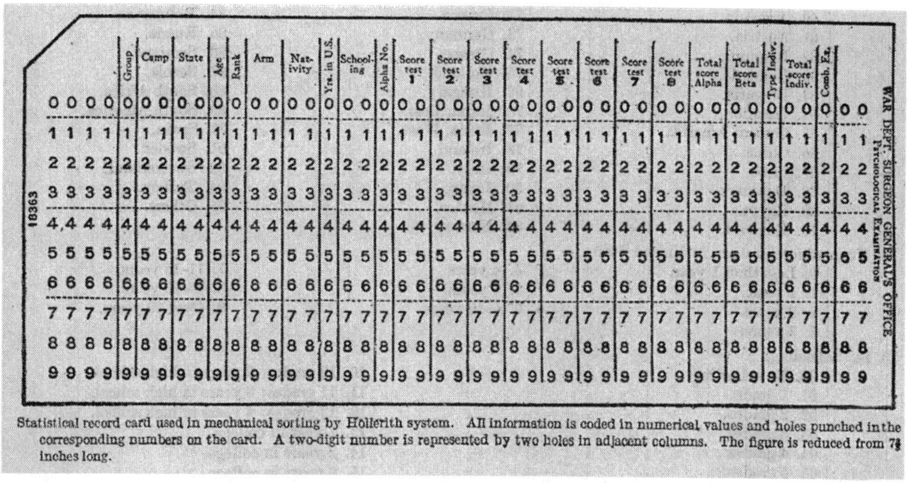

Statistical record card used in mechanical sorting by Hollerith system. All information is coded in numerical values and holes punched in the corresponding numbers on the card. A two-digit number is represented by two holes in adjacent columns. The figure is reduced from 7¼ inches long.

FIG. 2.1 Hollerith card design for sorting and analysis of a 160,000-card random sample from the 1917–18 army intelligence tests (Yerkes 1921, part 3, 538)

that "there are many mental processes which can not be grouped under the headings which have been considered so far, and it has been a custom for psychiatrists to deal with these mental states under the heading of general intelligence."[64] This remarkable attempt at a negative definition of intelligence was in fact rather unremarkable in the years in which intelligence testing was still emerging.

The stabilization of these tests, at least in the American context, soon occurred after the US Army's widespread use of intelligence testing on recruits in the first World War. Psychologist (and later famed primatologist) Robert Yerkes chaired the Committee on the Psychological Examination of Recruits. This group developed the Army Alpha and Beta intelligence tests that were administered to over a million recruits and draftees.[65] One of the contributors, Columbia's Robert Sessions Woodworth, soon after published a chapter on "Intelligence" in his 1921 textbook *Psychology: A Study of Mental Life.*[66] He there referred to the Binet tests as "extraordinarily useful," discussed the revisions by Terman (who also contributed work on the army intelligence tests), and argued that "the tests have been standardized by actual trial on large numbers of children, and so standardized that the average child of a given age can just barely pass the tests of that age."[67] This sort of claim was crucial for clinching the scientific status of the test as a source of objective evidence. Individuals newly subject to these tests might dismiss them as pencil-and-paper speculations. By attaching scientific authority to the tests, psychologists made plausible the idea that these informational instruments were telling people something true about themselves.

Woodworth sought to embody the scientific type of investigation necessary for the construction of objectivity in experimental apparatus. His 1932 autobiography candidly relays how the influence of William James at Harvard and G. Stanley Hall at Johns Hopkins made him feel able to "glimpse the frontier of scientific discovery" such that one night, in returning to his room after attending a lecture by Hall, "I inscribed a card with the motto, INVESTIGATION, and suspended it by my desk."[68] Woodworth recalled of his later research at Columbia under psychologist James McKeen Cattell, then editor of *Science*, that "his emphasis on quantitative experiments of the objective type, and his interest in tests for individual differences, were powerful influences," and noted also his study of anthropometry and statistics under anthropologist Franz Boas.[69]

When it came to intelligence tests, Woodworth was convinced of their objectivity. In his chapter on "Intelligence," he proposed to analyze "in what [intelligence] consists," and stated that "we can best proceed by reviewing the intelligence tests, and asking how it is that an individual succeeds in them."[70] Woodworth held that the tests could serve as a definition of intelligence itself such that analyzing intelligence meant reviewing factors for success at tests of intelligence.[71] The apparent circularity of this logic is in fact part and parcel of how objectivity in this field was ratcheted up out of nothing sturdier than thin air. Tests were tested against other tests, whose validity was checked by other tests. What is most remarkable in this is Woodworth's self-certifying invocation of the tests in a cultural milieu in which intelligence testing had assumed an at best unsteady place in manuals of mental measure less than a decade earlier.

Readers may sense a critical quip in my observations. But I want to be clear that I make no pretense of having the standing to dismiss, debunk, or deconstruct sturdy instruments of mental measure. What is more on point, I think, is to track the historical conditions that made it possible for such measures to become sturdy, durable, and lasting into their future (which is now our present). We can learn much from setting personality psychology's historical conditions of acceptability beside other historical conditions that index charged commitments. Consider, for example, that intelligence testing stabilized in part because of its work in helping to fulfill the aspirations of Jim Crow racisms and Know-Nothing nativisms. This influence has been traced by historian Stephen Jay Gould and genealogist Ladelle McWhorter, among others.[72] Their research demonstrates that one dimension along which we can trace the politics of psychological testing in the early twentieth century is their often-overt racism. Faced with this, we can stand immovably on the moral ground of denunciation. Or, we can go on to ask about these moral abominations what it was that may have made personality psychology so effective in a historical context where so many were pursuing older racisms by newer means. What were the mechanics of its effectiveness, and how did these come into contact with contemporary conditions of acceptability? My argument is that the adoption of an informatics apparatus was a key to the success of this field of inquiry. Narrating that adoption—and the arguments required to get it off the ground—is the focus in this chapter. A discussion of how such apparatus were leveraged in the context of then-transforming ideas of race is not, however, my focus here. To do justice to that topic, I need to reserve space for its story to be told on its own terms. I do so in the next chapter, where I discuss the informatics of race in the rather different context of real estate appraisal and residential segregation.

THE FIRST PERSONALITY TEST, CA. 1917

Intelligence testing took less than two decades to achieve a powerful epistemic authority that persists to this day. With intelligence tests, sciences of measure began colonizing domains previously preserved as sacrosanct. Our very minds became correlates of standards of rigorous informatics. The speculations of philosophers of mind were increasingly edged out by the new technicians of mental measure. The armchair, the treatise, and the sermon were set aside in favor of calipers, questionnaires, and regressions. If information could invade the mind's capacities for cognition, then perhaps it could do the same for seemingly even more elusive aspects of mentality.

An emblematic moment in the emergence of new mental measures was the production of the first personality test: the Personal Data Sheet. The US Army's primary focus in World War I psychological testing was on intelligence. Where these proved insufficient, however, further tests were solicited to measure qualities relevant to soldierly duty that could not be corralled under intelligence. The National Research Council thus appointed in 1917 a Committee on Emotional Fitness for Warfare. Its chair was Robert Sessions Woodworth. The committee developed a test for administration to large groups of recruits with the purpose of identifying those for whom further individual examinations would be needed. This would limit the resources needed for subjecting incoming soldiers to individual psychological examination. Woodworth's Personal Data Sheet—or Psychoneurotic Inventory, as it came to be called—was completed in 1918 as the war itself was coming to an end. It was only briefly administered before the armistice, yielding data on around a thousand new recruits and 274 already-diagnosed patients.[73]

I follow a slew of references in calling this "the first personality test." But what does that mean, the first "test" of personality? Had not abnormal psychologists been scrutinizing patients' emotional lives for decades? Yes, but that scrutinizing was hardly conceived as a "test" and was certainly not administered as one. One might as well conflate pharmacology and talking cure, since both are therapeutic. Analyses, clinical examinations, and long-term case studies are all instruments in the psychological armamentarium. But they should not be confused for the tests of the scientific psychologist whose research object is personality.

To develop the first personality test, the team examined extant accounts of mental illness to determine symptoms precipitating mental breakdowns.[74] These were then used to generate a set of yes-or-no questions that were tested against "normal" individuals so as to narrow the questions. The resulting schedule included 116 questions, with the assumption that normal persons would on average give about ten "wrong answers" (or indicating answers) to these questions. One early commentator noted that "any individual who answers 20 of the questions wrongly should be suspected of instability" and "if the number of 'wrong' answers is greater than 30 grave suspicion of abnormality is warranted."[75] Here is a sampling of the questions (with the abnormality-indicating answer in parentheses): "29. Have you lost your memory for a time? (Yes); 35. Were you shy with other boys? (Yes) [the test was intended to be administered only to males.]; 51. Have you hurt yourself by masturbation (self-abuse)? (Yes); 58. Are you ever bothered by the feeling that people are reading your thoughts? (Yes); 102. Did you ever have St. Vitus's dance? (Yes)."[76] A small number of the questions were understood to be unlikely to produce false positives:[77] "32. Were you considered a bad boy? (Yes); 50. Do you think you have hurt yourself by going too much with women? (Yes); 55. Did you ever have the habit of taking any form of 'dope'? (Yes); 83. Do you ever feel a strong desire to go and set fire to something? (Yes); 110. Has any of your family committed suicide? (Yes)."[78] With these questions, and the algorithmic analysis of the recruits' answers to them, personality began to be quantified.[79]

The pencil-and-paper questionnaires of the psychologists were remarkable little devices. They were compact and portable. They were designed to be self-administered (filled out by the test subject) with a minimum of (very precise) instruction. They did not require a detailed interview with a psychologist. They did not allow for special questions pertinent to an individual being tested. They did not demand any interpretation. They produced binary (yes or no, agree or disagree) or numerical (scaled) answers ready for computation.

Personal Data Sheet

Answer the Questions by underlining "Yes" when you mean yes, and by underlining "No" when you mean no. Try to answer every question.

Name ....

Residence ....

Age.................... Weight.................... Height....................

Race................ Nationality............................ Sex................

Education { Elementary School ....
           { Junior High School..
           { Regular High School...

Occupation.................................... Weekly wages..............

Date ....

1.	Do you like to play with other children?	YES	NO
2.	Would you rather play by yourself alone?	YES	NO
3.	Do other children let you play with them?	YES	NO
4.	Did you ever run away from home?	YES	NO
5.	Do you ever feel like running away from home?	YES	NO
6.	Do people find fault with you much?	YES	NO
7.	Do you think people like you as much as they do other people?	YES	NO
8.	Do you mind crossing a bridge over water?	YES	NO
9.	Do you mind going into a tunnel or subway?	YES	NO
10.	Are you afraid of water?	YES	NO
11.	Are you afraid during a thunder storm?	YES	NO
12.	Do you feel like jumping off when you are on a high place?	YES	NO
13.	Are you afraid of the dark?	YES	NO
14.	Are you often frightened in the middle of the night?	YES	NO
15.	Are you afraid of noises in the night?	YES	NO
16.	Are you troubled with dreams about your play?	YES	NO

FIG. 2.2 Page 1 of Robert Sessions Woodworth's "Personal Data Sheet" as later adapted for general use and published by C. H. Stoelting Co. in 1924, from the Psychology Tests Collection, Archives of the History of American Psychology, Drs. Nicholas and Dorothy Cummings Center for the History of Psychology, University of Akron; the text of the test questions (but not the original typesetting and formatting) is reprinted in PsycNET's PsycTESTS database as well as in Franz 1920, 171–76, and Hollingworth 1920, 120–26.

Their rhetorical style was technical and turgid rather than literary and vivid. Their results were drawn up in tables and delivered as percentages.[80] They became objective, repeatable, reliable experiments even if it was never quite clear for the first decades of researchers just what their object of study was. Looking back ten years later, Woodworth admitted that the results of the first personality test he devised "have never been striking," and yet the very idea "still seems to have possibilities of usefulness."[81]

Woodworth may have been playing coy here. He would have known how extensively the Psychoneurotic Inventory had been redeployed for nonmilitary use after the armistice, even if most of these later applications involved revisions. One of the first important revisions was by Ellen Matthews of Los Angeles City Schools, who published a one-hundred-item version of the test in a 1923 article in the *Journal of Delinquency* that featured extensive tabular data as well as hand-drawn graphs and hand-drawn bar charts.[82] A 1925 revision published as "A Mental Hygiene and Vocational Test" by Donald Laird at Colgate replaced the yes-or-no design with a graphical format in which test takers were asked to connect horizontal dots to indicate a range of magnitude over labels listed beneath the dots.[83] Laird's proposal offered an ingenious approach to personality measurement: have the test takers themselves give their answers in the form of continuous measure (eliminating the need to smooth binary replies into analog measures). A revision by Maxwell Papurt of Ohio State designed for use in a psychopathic hospital was published in 1930 accompanied by a hand-drawn scatter-plot graph mapping patients and their various diagnostic conditions revealed by the test.[84] These tests indicate both the early work of technical adjustment to Woodworth's inventory and the wide range of early sites of deployment from public schools to vocational testing to clinical psychology.

Soon a whole bevy of researchers began devising tests of their own to interrogate that elusive because not-yet-existent object of personality. Among the dozens of tests produced in the 1920s were the X-O test by Pressey and Pressey in 1920 (also known as the cross-out test), the Rorschach inkblot test in 1921, the will-temperament test by Downey in 1922, the personality rating scale developed by Marston at the Iowa Child Welfare Research Station in 1925, the ascendance-submission test by Allport in 1928 (of which more later), and the American Council of Education's 1929 "Personality Report."[85] By the end of the decade, personality testing was a stable method for scientific research of a stable object of inquiry: personality.

Back at the decade's dawn, a young Harvard graduate student named Gordon Allport wrote a survey of the fledgling field in an article that has been described as "the first American review of psychological literature on 'personality

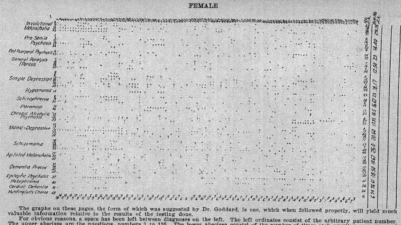

MALE

FEMALE

The graphs on these pages, the form of which was suggested by Dr. Goddard, is one, which when followed properly, will yield much valuable information relative to the results of the testing done.

For obvious reasons, a space has been left between diagnoses on the left. The left ordinates consist of the arbitrary patient number. The upper abscissa are the questions, numbers 1 to 116. The lower abscissa consist of the number of times each question was answered pathologically. When we turn to the vertical columns on the right, we find the first to be the number of questions answered pathologically by each patient whose number corresponds to that on the vertical left. The second vertical right column is the average number of pathological answers for each diagnosis.

For example, patient number 2, an Involutional Melancholia, answered questions 1, 9, 10, 11, 12, 13, 18, 20, 23, 24, 33, 41, 43, 58, 61, 68, 69, 77, 90, 94, 97, 99, and 100, a total of 23 answers, pathologically. However, we see that the average number of pathological answers for the Involutional Melancholia group is 29.25, therefore patient number 2 was a good deal under the norm for her group.

Or if we wish to determine how many times question 75, for example, was answered pathologically, and by which patients, we start at the upper abscissa and follow the vertical line to the lower abscissa. We then see that question 75 was answered pathologically by patients 4, 8, 9, 46, 19, 35, 30, and 40, a total of eight. By referring to the column on the left, the diagnosis of each of these patients can be seen. It is suggested that whenever necessary a ruler be used in reading this graph.

FIG. 2.3 Scatter-plot scores in Maxwell Papurt's revision of the "Personal Data Sheet" (Papurt 1930, 342–43)

and character.'"[86] Allport's review followed recent developments in carefully distinguishing personality (traits expressed in behavior) from character (evaluations "according to prevailing standards of conduct").[87] He also,
somewhat confusedly, included within the purview of personality a range
of theories and tests concerned with intelligence, emotion (under the heading of temperament), and volition.[88] What is of greatest interest in the article
is Allport's confidence in the tests as sources of accurate measures of still-
undetermined qualities: "In all probability, as has been the case with the study
of intelligence, we shall be able to give reliable quantitative results before we
understand the precise nature of that with which we are dealing."[89] Allport's
rhetoric here is quite casual, but the claim itself is stunning—for what he
ventured in that thought is that quantitative measure can proceed reliably
even as it precedes any other facts about what is being measured. Allport
was not unsettled by the risks of the venture: "Notwithstanding the dangers
and difficulties encountered in devising and employing rating scales, we are
forced to recognize this method as the only available objective criterion of
personality."[90] Perhaps early personality researchers exhibited the vice of excessive confidence. But that confidence was also expressive of the virtue of
relentless curiosity. Allport's view of the field is fairly summed up in another
1921 article he coauthored with his older brother Floyd: "At the start the investigator is only vaguely aware of the things he intends to measure, and he
can only guess at test problems and procedures which will indicate the traits
which he selects."[91] Guess we must, they implied.

A similar approach was proposed a few years later by Percival Symonds of
Columbia Teachers College in his taxonomic survey of "character measurement" test methods.[92] Commenting on the overall state of the surveys, Symonds exhibited keen confidence. He urged that progress in the field would
be expedited "if instead of trying to construct 'valid' tests we try to construct
'reliable' tests."[93] By this he meant to encourage work on tests that were consistent either internally or across different test environments, rather than
striving for the ideal of a test that could be matched against prior knowledge
of the test object. Symonds recognized the counterintuitive nature of dispens-
ing with objective reality in objective science. He even mockingly described
his own suggestion as the view that says, "I don't know what I am trying
to measure but I am trying to measure it accurately."[94] Despite this precious
self-awareness, he still held that, "once having constructed a reliable test, it is
comparatively easy to find out what it measures and all is gain."[95] The future
of psychometric researches would prove Symonds right in his prediction that
reliability can lead the way to validity. This approach would, however, open

up vigorous debates about the status of traits as either merely nominal entities or more robust realities.

In conducting such debates, early personality psychologists found themselves facing a strange feature of any standard of measure.[96] Once a unit is rigorously standardized and portable tools for its measurement are furnished for use, the unit measured by the standard *becomes real.* Did *real* inches exist before the ruler? What about *real* volts before the voltmeter? Metaphysicians may debate the ontological status of units in the absence of their measuring devices, but as soon as the technicians with their devices show up, the question loses all its seeming metaphysical mystery. At that point, a *real* inch is no longer a metaphysical conundrum to be debated in the parlor but a technical question to be settled by the ruler. *Is it real?* the metaphysician asks. *A real what?* the technician replies. *Do you mean a real inch?* they then ask.[97] If so, the question admits of an answer to anyone carrying a measuring device. If the technician is a personality psychologist, then their follow-up will be, *Do you mean a real personality trait?* If so, psychological apparatus can help produce an answer. Herein we find the wisdom in talk of "traits which *exist.*"[98]

Those who doubted the existence of traits would be confronted with a whole battery of psychological instruments producing traits as real correlates of measures. Grace Manson published a 1926 National Research Council itemization of 1,364 personality researches under the title *A Bibliography of the Analysis and Measurement of Human Personality Up to 1926.*[99] A clear majority of her items date from after the war. That same year, two researchers from Columbia Teachers College published a 196-item bibliography of "Personality and Character Tests."[100] A third bibliography published the following year listed 2,295 items over nearly two hundred pages.[101]

Such bibliographies evidence a rising cadre of science-minded researchers who were shaping personality psychology into an authoritative locale for formatting selfhood. What these researchers were producing was a crucial vector—just one of many, according to my broader argument in this book—along which could travel the informational productions of persons. Persons, possessed by a fascination with personality since the turn of the century, began to find themselves in possession of real personalities defined in terms of real traits capable of reliable measurement. How did those measures manage to muster the reality of what they measured? How did personality tests help solidify personality traits? The methodological questions with which Symonds and Allport were wrestling in the first years of the decade would prove to be one of the crucial issues on which would hang the scientificity of personality research.

MEASURING PERSONALITY TRAITS, 1927–1936

In 1921, still a Harvard graduate student, Allport was prepared to concede the forthright nominalism about traits as expressed by researchers like Symonds.[102] By 1928, now a member of the faculty at Dartmouth and on the cusp of a return to Harvard two years later, Allport was firmly rejecting the nominalist approach: "The concept of trait must first be established on rational, statistical, and if possible on neurological grounds, before it can be employed with justification."[103] In the seven years that separates these articles, Allport had come to realize the crucial role that the trait concept would have to play in personality psychology. Personality traits would need existential status if anyone were to take the science of personality seriously.

An entire technological ensemble of measure eventually made traits, along with the personalities they composed, real. Most important in this ensemble was the crucial psychological device of the test that would do so much to help make measurement reliable. Allport could then later claim in his field-defining 1937 book that "the extension of mental measurements into the field of personality is without doubt one of the outstanding events in American psychology during the twentieth century. . . . the swift output of ingenious tests has quite outstripped progress in criticism and in theory."[104] That this approach still orients the field's agenda today is evidenced by this claim in Larsen and Buss's 2010 personality psychology textbook: "More than any other approaches to personality, the trait approach relies on self-report questionnaires to measure personality. . . . Questionnaires are the most frequently used method for measuring traits."[105] What made the questionnaire test successful as an instrument of measure was its efficiency in algorithmically processing those data it was also efficient at recruiting. To see how such a compact informatics device achieved such epistemic authority in such a short period, we can follow the technical innovations of Allport on the cusp of his shift to a realist theory of traits.

Allport first stated his direct opposition to trait nominalism in his 1927 article "Concepts of Trait and Personality": "The truth of the matter is that neither measurement nor inventories of 'attitudes' and the like can be intelligible until the substantives themselves are clearly understood."[106] Allport's framing of the problem shows how crucial the trait concept had become for personality research: "If 'traits' are merely nominal entities, there is still the problem as to *what constitutes the existential unit of personality*. . . . For convenience it is proposed to rescue the term 'trait' from the confusion in which it is embedded and to apply it consistently to designate the unit sought."[107] After a survey of recent uses of the concept, Allport concluded by simply affirming "the recognition of 'trait' as the unit of personality."[108] His argument is entirely unconvincing, and yet his conclusion must have been utterly attractive to his readers. Interestingly, it also anticipates the precise path that personality psychology would pursue.

If the argument for trait realism in the 1927 article is empty, Allport hit on a more convincing approach the following year in an article titled "A Test for Ascendance-Submission" reporting on his just-published *A-S Reaction Study*.[109] Allport here played the role not of the theorist who would make an argument but rather that of the technician whose instrumentarium could be made to serve a theory. Allport in 1927 was arguing in the abstract that traits were the real units of personality; Allport in 1928 set argument to the side and simply produced an instrument for measuring traits of ascendance and submission. By locating the concept pair of personality-and-trait through algorithmic technologies of measurement, Allport effected a conclusion that did not so much stand in need of argumentative justification as it enacted a quiet substantiation. I analyze this article in detail because magnifying its mechanics clarifies how the kind of algorithmic analysis on which personality psychology came to depend was an enactment of a raft of formatting techniques that effectively fastened persons who could then be said to have personality traits.

Before turning to this analysis, however, I want to preface my discussion with a note about recent work on the social dimensions of algorithms, for this scholarship often treats algorithms as technologies whose social relevance is new. By excavating algorithms as information-processing technologies embedded in wider assemblies of formatting, my approach opens up the possibility of placing a historical check on this tendency of emphasizing the "increasing" relevance of algorithms without assessing the past reach of algorithms against which their greater relevance "now" might be measured.[110] The significance of algorithms can be recent only if we peel them off from the broader data assemblies in which they are put into operation. Looking more widely, however, we can recognize that algorithms have been helping to format our very selves (and so much more) for nearly a century. As such, it is not some supposed recent expansion in algorithmic deployment or capacity that deserves our critical attention, but rather a long-entrenched disposition to algorithms that has well prepared us to regard ourselves as subjects of algorithms.

Consider the motley assembly through which Allport's article characterized the technology of the test: numerous hand-drawn charts, a decile table on which is grouped test scores, the claim that "the trait, statistically considered, falls into a normal distribution," the suggestion that "we are dealing with a relatively homogeneous statistical phenomenon susceptible of measurement and scaling,"[111] a sampling of the test questionnaire (consisting of verbal descriptions of situations to which the test taker responds by indicating how he or she would act), a sampling from the instructions used to administer the test, a discussion of the process through which answers to the test questions were rated on the ascendance-submission scale and then converted into quantitative data, and, finally, the conclusion that "these traits have been found to be fairly constant characteristics of behavior, and to be measurable with a device which employs verbally presented situations."[112] With all of this, was Allport simply reproducing Symonds's proposal? Was he nominalizing traits as "normal," "constant," and "measurable" effects of inquiry? Despite possible appearances to that effect, there is one crucial difference. Symonds held that reliable measure would pave the way to later objectivity such that traits in the interim would be merely nominal. Allport was developing the view that the objectivity reality of traits could be definable just in terms of measure. Traits became like inches or volts: measure them and they are really there.

Allport's test and article are together representative of the measurement-centered approach the field would fortify over the next decade. Consider just three features of informationalization in his approach, one from each of the three informational dimensions of input, processing, and output. I begin with, and consider at greatest length, the informatics of algorithmic processing, as it was the most important dimension in this context for securing scientificity.

The *A-S Reaction Study* performed its work of formatting most centrally through algorithmic data processing. In this, it is exemplary of the approach that personality research adopted over the next decade. Having solicited input from test takers on a self-reporting questionnaire, the test would initiate its work of algorithmic processing. This data processing involved a relatively simple set of procedures, a self-contained rule set that anyone with basic arithmetical skill could follow.

In considering the test's algorithms, it is crucial to register the non-algorithmic research that went into their production. Doing so gives specificity to a general feature of all algorithms: there are no algorithms that are themselves the pure products of algorithmic work. Design decisions must always be made. Allport notes in a section of the article devoted to describing the construction of the test that an initial version was administered to six hundred college students. Those data were used as the basis for computing raw averages, then simplifying the averages for each question by computing its deviation from the overall average, then computing by correlation the advantage of retaining two decimal places in the computation, and, finally, computing a table of norms (normal scores across those six hundred test takers) that would be used as a basis against which future individual test takers' summed scores could be compared.[113] All this work of computing and refining averages required meticulous intervention, or what we could call design decision.

On the basis of numerous such design decisions, the research process was eventually made to yield a finalized test that could compute personality ascendance or submission by way of a relatively simple algorithm. Once data was obtained by administering the questionnaire, the information-processing phase would begin. The test taker (or someone administering the test) would follow a basic three-step algorithmic procedure.

First, the test manual included a lookup table for converting the obtained data, check marks beside blanks, into numerical quantities. Consider, as an example, this question: "Have you crossed the street to avoid meeting some person? Frequently _____, occasionally _____, never _____." Respondents checking "frequently" or "occasionally" are given a score of –2 for that question, while those checking "never" are given a score of +2 (minus scores indicate submissiveness and plus scores indicate ascendance).[114] Note that these are the score values attributed to men taking the test, but for women taking the test, checking "frequently" yields –2, checking "occasionally" yields a value of zero, and checking "never" yields a score of +1.[115] This division of personality along the lines of sexual difference is just one of countless algorithmic formats built into the test. That noted, my point here is *not* one that depends on how identity categories are loaded into the test. Even a revision of the test that attended more carefully to formats of sex and gender would still perform the work of formatting, for it would convert into measurable numbers the checkmarks elicited from test takers by plain language questions. A check mark on a form beside an ordinary English phrase in response to an ordinary English question is transformed into a numerical value, even though neither the phrase nor the question nor the check mark are themselves—or even reference in the least—numerical values. Today, such conversions to number are ubiquitous in testing of all types. Already in 1928 Allport did not condescend to argue for this kind of formatting work. He simply performed it.

Second, with quantified responses in hand, the test could commence with its next step in processing. This was also simple: an arithmetical computation of an overall score. All the plus, minus, and zero values yielded by the quantification of the test taker's answers were summed.

Third, and finally, the computed sums could be visually compared to a chart representing the subject's "degree of ascendance-submission."[116] There were located the meanings attributed to the scores. These meanings were also products of the non-algorithmic processing work of the research and design team. These results were thus both a product of an automated algorithm and an effect of a craft of test design.

Beyond the formatting work of algorithms, the tests were also information-intensive in their inputs and outputs. As to their data outputs, one notable feature of Allport's article is its hand-drawn charts—which, to today's eye trained in a visual culture of sharp angles and straight lines composing plots and charts, look enormously rudimentary. Indeed, in the article itself, the axes and their labels are not typeset but hand-drawn and handwritten. The weights of the plotted lines are thus inconsistent, and one of the charts is even reproduced on the page at a noticeable skew from the alignment of the printed text. What makes these charts nonetheless impressive is that they convey the stability of information despite their dirty appearance. Later charts would look cleaner once printers began to develop ways of improving the visual presentation of informational displays that at an earlier time had no alternative but to be drawn by hand. Yet even early hand-drawn data could not fail to be convincing.

A final informationalizing feature of the test is found in how it handled test data as inputs that could be efficiently prepared for processing. The A-S Reaction Study procured its data on a simple form whose style is ubiquitous in the genre of early personality research. At the top of the form the test taker indicates his or her name and age. A blank is left for the score that the subject (or perhaps the person administering the test) can fill in later. The test itself consists of a series of questions. Each question concisely describes a scenario and then offers from two to four possible responses. Test takers may not write their own answers. They are instructed to respond by "checking the answer which most nearly represents your usual reaction."[117] This is quintessential formatting technology. However one might actually respond to the situation described by the question, the response one must give in the test is already precisely formatted. The subject of the test is thereby fastened to formats by virtue of which their personality is made scientifically legible.

Having surveyed three kinds of formatting performed by the A-S Reaction Study, I want to distance my analysis above from the suggestion that these formats are unjustifiable tricks by which researchers mathematically reduce more complex qualitative phenomena. Such a critical gesture of denunciation is not my intent. Rather, I aim to interrogate the historical conditions of possibility whereby present truths became true. I leave to others to contest whether we *ought* to be fastened by the results of this or that test. Whatever those debates would yield, it is of decided interest *how* we have come to be fastened by the technical formats embedded in all our tests. For the historical conditions of our fastening ought to be of value to any effort to contest or cement the particularities of our being fastened in one way rather than another.

rounded to one place decimal with ratings. A random selection of 100 cases was used, with the following result:

$$r_1 = + .599 \pm .043$$
$$r_2 = + .586 \pm .045$$
$$r_1 - r_2 = \quad .013$$
$$PE\,(r_1 - r_2) = \quad .061$$

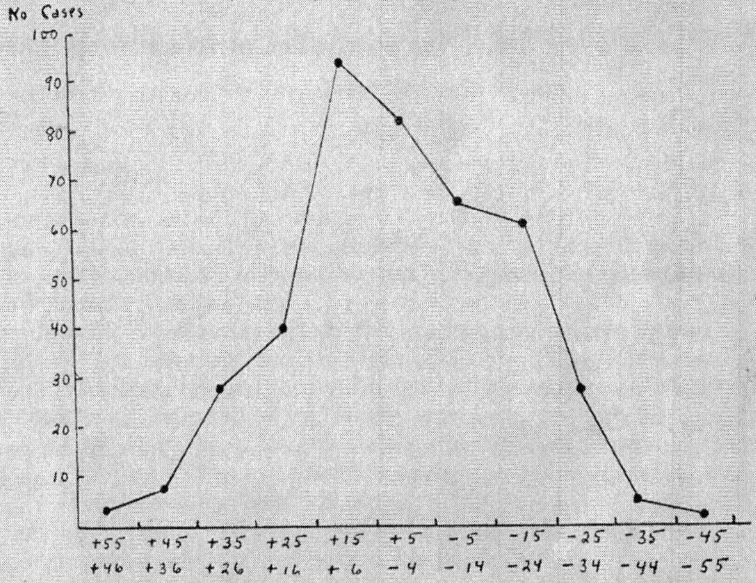

FIG. 2

DISTRIBUTION OF SCORES (MEN)

Four hundred cases (the original standardization group at Dartmouth). The intervals on the base-line represent scores grouped into intervals of ten. In a subsequent standardization group (N=327) two cases higher than +55 were found.

Thus, the slightly greater validity obtained for the test by using two place decimal scores was less than the probable error of the difference. The one place score was adopted, and the decimal omitted.

5. The total score for the subject on the test as a whole is obtained by adding together all of the ascendant scores (plusses)

FIG. 2.4 Page portion from a Gordon Allport article showing the tilt of the hand-drawn graph relative to typeset text (Allport 1928, 129)

One such historical condition that I have suggested is crucial for the fastening work of personality psychology is the early theoretical controversy over the existential status of traits. Though this was a theoretical controversy, it was not resolved by means of theoretical argumentation. It was, rather, resolved technically. On one side of the debate was a theoretical ontology of what can count as real. On the other side of the debate were algorithms that simply

formatted data such that it came to be real enough. On one side was the ontological category of "validity"; on the other was operative technical "reliability."
The changes in Allport's thinking across the 1920s exemplify this shift.

In 1921, Allport had worried that reliability without validity would lead
to nominalism. By 1928 and the development of his ascendance-submission
test, Allport was ready to regard reliability as a proxy for the reality of what is
measured: "The reliabilities, approximate though they are, seem to indicate
that the test measures with fair consistency some constant factor in personality. They are, in short, evidence for the existence of the traits defined earlier
in this article."[118] The shift here is quiet but crucial. Allport had once worried
with Symonds that traits might not be real if we cannot find any independent
criteria of their existence beyond testing apparatus. But later he was simply
unconcerned with that possibility and could endorse traits as real on the basis
of the tests alone. If this shift seems remarkable, consider now the sentence
following the lines quoted just above: "It is not easy to determine the validity
of such a test as the one here offered."[119] The previous line had just asserted
that existence is a function of reliability. What work, then, might even be left
for an independent concept of validity? Allport in fact gave it precious little
role to play and was content to leave validity as "uncertain."[120] With reliable
measurement taken as sufficient for existence, there was little need to trou-
ble with validity beyond the decorum of nodding in its direction. Allport's
mature position, which he would retrospectively dub "heuristic realism" in
1966, was in fact not all that different from his earlier 1921 view shared with
Symonds.[121] At both ends of the decade, Allport held that reliability was the
crucial scientific achievement such that validity could be left to the side. The
difference was that Allport was now pleased to call "existent" whatever was
being reliably measured. To a critic this could appear to be a massive sleight
of hand. But such a criticism begs the question. I decline to criticize the view
because what matters for my purposes is not so much the validity of Allport's
conversions as the complex processes through which they were made into
reliable techniques for recomposing selfhood. I too am a reliabilist about the
so-called reality of that which our data bodies forth.

By the end of the 1920s, Allport solidified a psychometrics of personality.
This set the stage for a decade of work in the 1930s, when his most influential
contributions were published. His first major article after returning to Harvard in 1930 was "The Field of Personality." This fifty-four-page review offers
a programmatic statement of personality as a still-maturing field of research.
It notes that "interest in the subject is largely a post-war phenomenon, but
though recent it has already reached astonishing proportions."[122] Allport and
Philip Vernon imposed some order on the field by classifying the extant litera-

ture under six headings. In the discussion of measurement under the heading of "Experimental Methods," Allport again confronted the problem of reliability and validity.[123] He conceded that "most of the literature on measurement contributes little to the theory of the topic," such that "a general critique of testing methods and results is badly needed."[124] Allport and Vernon focused on "many recent advances" that they described as beginning in 1925.[125] Foremost is the fledgling technique of "validation by sampling, now known as theoretical validation or internal consistency."[126] The central idea is that of validating results as part of a "composite battery" of inquiries rather than validating them against some untestable but purportedly obdurate object of inquiry.[127] This could take the form of a testing instrument employing multiple test items or multiple testing techniques demonstrated to be consistent with one another: "If the test or the composite battery is internally consistent, then it is considered as valid whether or not it agrees with other criteria such as ratings."[128] Allport's ascendance-submission test of 1928 was cited as an example of an instrument that works by self-correlation.[129] Through the promise of self-correlating validation, the test technology could close its own loop. In the immediately following section, Allport and Vernon discussed "Consistency and Trait" in such a way as to leverage reliability toward the conclusion that "a justifiable statistical procedure" shows "promising" levels of intercorrelations, especially in tests where "experimental situations are more natural so that the testees' true consistency and normal traits are able to emerge, undistorted."[130] Thus was Symonds—referred to as "a doughty opponent" in the course of discussion—definitively discharged.[131]

The following year, Allport offered a programmatic account of traits in his January 1931 article "What Is a Trait of Personality?"[132] Three decades later Allport would describe this article as "probably the first attempt to formulate what has come to be called 'trait theory.'"[133] The program consisted of an eight-plank platform. Plank number one: "A trait has more than nominal existence."[134] Plank number four: "The existence of a trait may be established . . . statistically."[135] Traits are real. How do we know? Because they are measurable by informatics apparatus. What made this argument work was not a sophisticated theoretical ontology but the deployment of new techniques then gaining acceptance as capable of showing—that is, making manifest—reality. Constituting an entity as an enrollee in a qualifying practice is equivalent to dragging that entity helplessly into reality. There is no point in arguing over the metaphysical-cum-ontological status of these entities in the face of instruments that can reliably measure different quantities of said entity. "Whatever exists at all exists in some amount," E. L. Thorndike had written in 1918, approvingly quoted by Allport in 1930.[136] Thorndike, Allport, and their interlocutors were writing, quoting, and inquiring in that great Galtonian spirit of measure.

Having articulated in his 1931 article why and how traits are real in the appropriate sense, and having demonstrated a kind of technical accomplishment capable of giving evidence for such a claim in his 1928 test, Allport would over the course of the next decade produce an array of additional evidence for his view in the form of test after test in which traits could not but be real objects of measurement.[137]

Among these was perhaps the crowning achievement of this period of his work: his 1936 "Trait-Names: A Psycho-lexical Study."[138] Here Allport and collaborator Henry Odbert provided a major contribution to psycho-lexical personality informatics by combing through the 1925 edition of *Webster's New International* to craft a list of 17,953 English-language trait names classified into four categories. The 176-page monograph is an overwhelming technical accomplishment. Reading these four long lists of trait names would be a feat in itself, let alone the survey of the dictionary that funded their production. Allport and Odbert rightly noted that "the naming of traits is, then, a perilous matter."[139] Part A of the article opened with a twenty page review of arguments between nominalists and realists about traits, concluding that "the theory we present then holds that trait-names are symbols socially devised (from a mixture of ethical, cultural and psychological interests) for the naming and evaluation of human qualities."[140] This casual conclusion suggests that they felt confident enough in the matter being settled that they could get on to what the article really wanted to establish. Part B was a fifteen-page discussion of methodology that opens with reference to Galton.[141] Finally, part C offered a 132-page list of trait names. The list is not an argument for trait realism but rather an instantiation of such realism in the form of an informatics device that is today a standard tool for research kits in numerous disciplines: a lexical database.

PERSONALITY INFORMATICS STABILIZED, CA. 1937

Allport's methodology and rhetoric of psychology as an informatics science was finally systematized in his now-classic 1937 book *Personality: A Psychological Interpretation*.[142] The book calls attention on its first page to "a new movement within psychological science" that has made "notable" progress "within the past fifteen years."[143] *Personality* is thus presented as "a guide book that will *define* the new field of study."[144] Danziger notes that, "in the American Psychological Association Yearbook for 1918, not a single psychologist listed 'personality' as a research interest, but by 1937, 7 percent did so."[145]

Allport was perfectly perched at that moment when personality psychology came into being begging for definition.

Allport was explicit that he "sought above all else to respect the many-sidedness of the subject-matter of this new science."[146] One of his early chapters offers an exquisitely detailed history of conceptions of personality. It is of great historical interest for how a personality psychologist in 1937 saw the history of his domain. Allport's professional predilection for visualized information inflects his history in that it culminates at chapter's end in a fifty-node network diagram of the historical concepts of personality surveyed in the preceding pages. The diagram makes clear the importance ascribed therein to specifically psychological definitions, which are visually represented as the synthetic culmination of two thousand years of personality theory across domains as disparate as law, ethics, poetry, and theology. For psychology alone is able to pull together a coherent concept out of the scatter of the ages. Allport proposed this as his psychological definition: "Personality is the dynamic organization within the individual of those psychophysical systems that determine his unique adjustments to his environment."[147] This is node 50 on the diagram, at the bottom in the culminating position. Just above it is node 49, which is Allport's more "terse" definition that is "acceptable enough in principle" even if "brief and vague"; the content of that definition is: "What a man really is."[148]

Allport's focus throughout *Personality* was the scientific status of psychological approaches to personality. This appears, for instance, in his discussion—immediately following his definitions—of why character and personality are nonequivalent concepts. In a favorite phrase, Allport claimed that "character is personality evaluated, and personality is character devaluated."[149] Invoking a strong split between value and fact, Allport insulated science against morality so as to combat the frequent "confusion of psychology with ethics."[150] Allport's final verdict confirmed the clear consequence of the early twentieth-century cultural shift from character ethics to personality psychology: "Character is an unnecessary concept for psychology."[151]

Chief on the agenda for establishing psychology as a descriptive science is its search for basic units. Allport held that "the progress of any science, it is said, depends in large part upon its ability to identify *elements* which, in the combination found in nature, constitute the phenomenon that the science has set out to examine."[152] He compared psychology to the paradigmatic sciences of chemistry (with its periodic table), biology (with its cells), and physics (with its quanta).[153] This sort of comparison, Allport noted again in an essay published the next year, "accounts for [psychology's] obsession to reach the eminence of scientific respectability."[154] Such eminence was in sight right from the preface

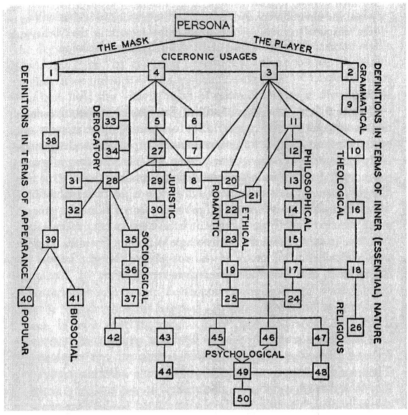

FIG. 2.5 Gordon Allport's diagrammatic representation of fifty definitions of personality from antiquity to his present (Allport 1937, 49)

to *Personality*, where Allport boldly stated that "the chief novelty of my own position lies [in the concept of] *traits*."[155] Traits were the basic elements of the science of personality psychology.

As to the reality of these units, Allport held that traits are "really there,"[156] and he did so on the basis of the methods by which psychologists can measure them.[157] Allport here finally concluded his earlier engagements with the realism-nominalism debate by asserting the reality of traits against the "specificists" (forerunners of today's situationists[158]) who argued that traits were always highly specific because highly dependent on context for activation.[159] Allport could now stand on the platform of almost a decade's worth of scientific work on traits that collectively established reliable instruments of measure. Indeed, the scientific status of the trait concept was so stable by

1937 that Allport could even engage in internecine debates over how to best conceptualize traits. Nothing marks a respectable field of inquiry like professional disagreement.

At the time, traits tended to be viewed in terms of their edges as defined by limiting cases. This creates complications when dealing with phenomena that have a tendency to be both fuzzy and quantitative. Rather than sort out these complications, Allport suggested shifting to a view of traits in terms of their foci. What he proposed was a then-novel definition of traits in terms of what he called "focalized dispositions."[160] If seen in this way, there could be no conceptual problem in traits being fuzzy at their edges—for fuzzy edges do not entail the absence of a central focus. Allport's suggestion was that traits should be analyzed not so much as "exclusive units" or "independent *factors*" but as "inter-dependent *traits*."[161] As he put it, "To seek *intelligible* units is a better psychological goal than to seek *independent* units."[162]

Allport was here criticizing the factor-analytic approach to traits that was gaining much attention in his day and later came to dominate the field. This method involves a statistical assessment of personality properties that cluster together in an effort to reduce the number of assessable properties to the smallest number of independent variables. The idea is that if a number of properties tend to cluster together with sufficient frequency, then that clustering can be redefined as a single property by resolving it into a single independent unit. This both allows more efficient treatment of a complex of properties and at the same time enables specification of the primary independent factors operative in such a complex. A list of properties that cannot be further resolved can be taken as a schedule of fundamental traits. Each item on the schedule would be irreducible and independent. This approach advantageously yields a schedule of elementary units that resembles the units of more paradigmatic natural sciences.[163]

Recall that Allport and Odbert had developed an exhaustive list of 17,953 trait names. The idea behind factor analysis is that we can and should reduce such an unwieldy list to a more manageable set, and possibly even to a minimal set of fundamental traits. Not long after Allport, Cattell in 1949 developed a sixteen-item model of fundamental personality traits.[164] In 1976 Eysenck and Eysenck settled on a three-factor selection.[165] Contemporary personality psychologists have reached consensus on a Big Five factor model, though there is some debate about adding an additional Big Sixth factor.[166]

Always troubling for Allport about any such schedule was the presumption that it carried advantages beyond mathematical simplicity. Allport criticized the factor-analytic method for its "assumption that *independent* factors are the desideratum of any theory of elements."[167] Occamites might favor factor analysis, but is what is simpler always what is needed? Allport cautioned against the "excessive empiricism" of factor analysis.[168] Later he fashioned a more lively phrase: "galloping empiricism."[169] Allport's view was that traits of personality form patterns of interaction so dense that to isolate independent factors was a hopeless goal: "So interwoven is the fabric of personality that it seems almost impossible to think of any patterns that are wholly unrelated to others."[170] One can always statistically separate off portions of the fabric, but in so doing we ought to have a specified purpose in view rather than is-suing a statement writ large about what is always fundamental in personality. Allport, then, recognized some value in factor analysis. His brief was only that it needed to be taken up pluralistically, without the insistence that iso-lated factors are independent in every case and so somehow fundamental—for sometimes the researcher might be interested in personality traits that would be not common across research subjects but idiosyncratic in particular individuals.

A pluralistic notion of scientific research was central to Allport's psychol-ogy. He would not allow the successes of statistical techniques—including his own successful deployments of these techniques—to rule out other ap-proaches to personality. "Personality is so complex a thing," he claimed, "that every legitimate method must be employed in its study."[171] The book's discus-sion of methods opens with "a graphic tabulation of the major methods em-ployed in psychological investigations of personality."[172] Radiating outward from the central operations of observation and interpretation are some fifty-two scientific methods for the analysis of personality.

Among the many methods whose values are affirmed are psycho-lexical analytics (method no. 4), the study of documentary sources (no. 10) and of personal documents (nos. 16–19), studies in expressive style (no. 24), scoring scales (no. 26), standardized psychometric scales (no. 29), laboratory experi-ments (nos. 37–38), and synthetic case studies (no. 52). Allport thus offered a capacious personality psychology, endorsing a type of pluralism that he would later make explicit in his essay "The Productive Paradoxes of William James."[173]

Allport's Jamesian pluralism was persuasive. Yet at the same time it is im-portant to recognize that it was a restricted pluralism, as all pluralisms al-ways are. He ruled out "charlatan characterologies (astrology, numerology, palmistry, and cranioscopy)" and uncritical everyday methods like "hearsay, prejudiced observation, [and] impressive coincidence."[174] Of particular inter-est are Allport's criticisms of psychoanalytic methods (nos. 41–46). His plu-ralism may have seemed to extend an accommodation to these approaches in that he offers no critical comments whatsoever on free association, dream

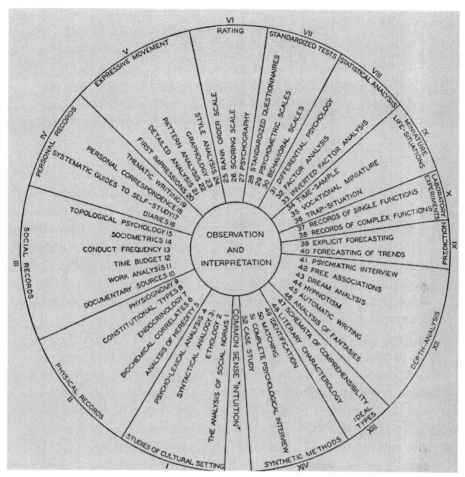

FIG. 2.6 Gordon Allport's diagrammatic representation of fifty-two methods for the study of personality arranged as spokes leading out from the central work of "observation and interpretation" (Allport 1937, 370)

interpretation, hypnotism, or automatic writing. These are, however, the only four methods in the chapter that were named but not discussed.

Silent for one chapter about psychoanalysis, Allport was elsewhere in the book cacophonously critical.[175] His criticisms almost always pertained to psychoanalysis's pretenses at "depth." In a retrospective autobiographical piece from 1968, Allport quipped that "depth psychology, for all its merits, may plunge too deep."[176] He cited there his own earlier criticism of Freud as his most frequently reprinted article.[177] In it Allport mocked analysts who think that their methods "tap the deepest layers of structure and functioning . . .

[when] most dynamic motives are more accurately tapped by direct methods."[178] The central problem with Freud's conceit of depth is that it results in a
"one-sidedness in his theory" according to which "motivation resided in the
id" and entirely within that deep place.[179]

Why was psychoanalysis such a foil for the otherwise pluralistic Allport?
His antipathy can be illuminated by the previous discussion of his criticisms
of factor analysis. In that case Allport's concern was with the monotonic
approach to personality implied in any nomothetic account of "fundamental" traits. Such approaches fail to allow for the possibility of alternative approaches. Allport, an avowed pluralist by contrast, did not so much seek to
delegitimate factor analysis as he sought to cut it down to size, affirming its
steps forward despite its claims to galloping success. Allport's criticisms of psychoanalysis should be seen similarly. Consider this comment on psychosexu-
ality: "Psychoanalysis, especially the Freudian variety, succeeds in the almost
impossible task of *over*emphasizing the role of sexual motivation and interest
in the human person. This is no small accomplishment, for—in Western culture at least—sexual tensions are in fact the most important single factor in the
development of most personalities . . . or rather *sex would be the most important single factor if there were any single factors, which there are not.*"[180] Just as
traits cannot be statistically reduced to independent functioning, neither can
sexuality be whittled down into a singular thing. For, Allport continued, "sex
in normal lives never stands alone. . . . It is indeed pervasive, but it is no longer
mere sexuality."[181]

A surface read might lead one to think that Allport's dissent from Freud
was over psychoanalysis's garrulous exhibition of sexuality. But that is not quite
right. The crucial difference concerned techniques. Psychoanalysis would only
ever interpret psychological phenomena through the supposedly fundamental
category of sexuality, whereas Allport objected to any version of the idea that
psychology could operate by wielding a single explanans that was always more
fundamental than every potential explanandum. Allport's pluralism could
thereby drive a wedge into a totalizing interpretivism and, in so doing, open a
path to the use of data within what may have been thought to be the preserve
of psychoanalysis.[182]

LEVERAGING INFORMATICS FOR CASE STUDIES, 1937–1942

When he published *Personality*, Allport was accomplished in statistical technology and the display of visual information. In the 1930 survey of "The
Field of Personality," he and Vernon had specified four classes of theories of

personality types. These were the a priori, biologistic, statistical, and qualitative (e.g., Gestalt) classes.[183] They confidently championed "the statistical
approach."[184] Yet seven years later, in *Personality*, he sounded in the preface to
the book a note of open caution about reliance on "currently popular statistical methodologies."[185] Readers familiar with Allport's own statistical achievement may have been surprised.

Equally striking was his celebration in *Personality* of "the case study" as
"the most complete and most synthetic of all methods available for the study
of personality," and "the most comprehensive of all [methods]" in which "the
psychologist can place all his observations gathered by other methods."[186]
How was the individual case study going to fit with algorithmic and statistical
techniques of trait analysis? Was this attempted combination a productive
paradox or a confused contradiction?

A key to Allport's enthusiasm for the case study as a valuable tool for
scientific inquiry into personality can be found in his antipathy to psychoanalysis. Rather than a departure from the psychological informatics for
which his own studies were paradigms, Allport's adoption of the case study
was a recruitment of a favored clinical method to a growing team of informatics techniques. Allport shrewdly recognized that he was historically
perched at a moment of transition in the field in which its former reliance on
individualizing case studies was giving way to the reliability of group data:
perhaps, then, the values of the former could be recruited to the achieve-
ments of the latter.[187]

Consider this earlier remark by Allport: "Until recently psychoanalysis
was the only system that held personality as a central concept, but it is evident that rival theories are in the making. Even though the formulation of
really adequate and consistent theories of personality is bound to lag behind
special research, this research is now so flourishing that the outlook for a
future systematic psychology of personality is bright indeed."[188] Allport now
sought to not only rival psychoanalysis, but also to beat it on its own terrain
of the case study. This sort of recruitment is paradigmatic of how scientific
informatics in those decades began to edge out hermeneutic approaches in
numerous fields.

For Allport, scientific method just meant analytic informatics: "As used in
this chapter the term 'method' has referred principally to ways of obtaining
information, *i.e.*, to the treatment of data."[189] That explains why his chapter
on methods was silent about psychoanalysis. Psychoanalysis will not traffic
in informatics. On Allport's definition, it therefore lacks method itself. Thus
we can understand Allport as enrolling the case study in informatics so as to
disinter it from a methodless hermeneutics. In his discussion of the genre of

case studies in *Personality*, he recognized the monumental task before him: "So little progress has been made in respect to this one most comprehensive technique that special attention is here given to it."[190] He offers a series of rules for dealing with case studies: maintain fidelity, deal with a known personality, write objectively and directly, describe the present personality, use both general and specific descriptions, take into account both formative influences and future plans.[191] The case study, he claims, "can include data drawn from tests, experiments, psychographs, depth-analysis, and statistics" such that it "embraces both the scientific (inferential) and the intuitive aspects of understanding."[192]

Without a detailed case of a case study, however, Allport would not get all that far in his recruitment. As Ian Nicholson notes, "Allport was more comfortable with the *idea* of the case study than its actual implementation."[193] So far as the record seems to reveal, Allport had never actually conducted a case study; at least, not in the usual sense in which that method is understood, by the time he wrote *Personality*.[194] Allport's most extended discussion of the case study approach came later, in a little-discussed book published by the Social Science Research Council in 1942 under the title *The Use of Personal Documents in Psychological Science*.[195]

The book opens by noting a recent surge of interest in "the personal document."[196] Allport defines "personal documents" or "subjective records" as "any self-revealing record that intentionally or unintentionally yields information regarding the structure, dynamics, and functioning of the author's mental life."[197] Examples include diaries, letters, autobiographies, and questionnaires. Allport intended the definition (by way of the "self-revealing" clause) to include first-person accounts but to exclude third-person accounts. Such "social documents" were institutional records and "the records of social agencies," transcripts of various kinds, medical records, newspaper accounts, and psychograms.[198] Both kinds of documents could have featured in his topic of interest, what he called "*case study* materials."[199] For both types of documents have in common that they "deal with *the single case*."[200] But third-person documents, known then and now as "paperwork" and "personal records," were not under Allport's survey. He expressed the hope that his discussion of "the role that the first-person document plays in psychological science" would turn out to "have validated in essence the use of third-person documents as well."[201] Yet those documents still await their processing by tomorrow's scientists of personality psychology.

Allport embarked on his fenced-off field of inquiry with an opening chapter on the history of first-personal documents as psychological evidence. He discussed early origins in Goethe, Wundt, Galton, Krafft-Ebing, and Lombroso before moving on to more detailed uses by James, Freud, and others. He concluded this survey noting that "not one single critical article on the *method* of the case study was published, and nothing beyond the most incidental and pedestrian use of *personal documents* is to be found."[202] Recall again my thesis that Allport was recruiting the case study for the purposes of scientific legitimacy. Some might think that more literary uses of personal documents are appropriate. Allport did not deny that; he only insisted that more scientific uses are also possible.

The turning point for the methodical use of such documents, Allport urged, was W. I. Thomas and Florian Znaniecki's 1920 study *The Polish Peasant*.[203] Following on that, and of greater import, was Herbert Blumer's 1938 critique of that book prepared for the Social Science Research Council.[204] Though Allport wrote of both personal documents and social documents in *Personality*, he did not there cite the then-seventeen-year-old study by Thomas and Znaniecki.[205] It seems likely that the book came to his attention by way of Blumer. Indeed, what held Allport's attention was an analytical problem flagged by Blumer: "Th[e] process of interaction between theory and inductive material which Blumer finds ambiguous is the essence of the methodological problem of personal documents."[206] The crucial issue, in other words, was that of the possibility of a scientific use of personal documents.

Personal documents, Allport claimed, can yield material admitting of statistical analysis.[207] He mentions "interrelations of items" as computed by a "Hollerith sorting machine" used in a colleague's analyses of student autobiographies.[208] Those surveys used punch card tabulators to compute results and thus represented an early foray into personality computation, where trait informatics would soon head.[209] Later in the book, Allport mused on one path that such a psychological informatics might eventually take, and entertained the speculative possibility that "a Hollerith machine worked on the basis of known frequencies by a robot could predict future behavior as well as a sensitive judge."[210] But he raised the possibility only to dismiss it: his reply to his hypothetical "code-and-frequency device" was that it must fail to engage in "the perceiving of relations."[211] It could approximate a passive intellect, but would not aspire to the "*intellectus agens* posited by Leibnitz [and] Whitehead."[212] Yet if all that separates Allport's informatics from information-theoretic computerization is a faint invocation of Leibniz, then the distinction is one that was bound to grow brittle before eventually becoming invisible. It was indeed already brittle, which is perhaps why Allport sensed the need to fend it off. Consider Allport's own discussion of the scientific study of a specific personal document he takes under his survey: an agoraphobic's autobiography.[213] His proposals include: pooling of diagnostic judgments, predictions for the future course of life, postdictions of causative or at least antecedent events, therapeutic experiments, study of the validity of the documents vis-à-vis independent sources, and a study of associational linkages.[214]

Confessions have long functioned as a technology for telling the truth about the self.[215] At some point, the confessional strategy came to be construed, by both confessors and auditors, in terms of texts whose hermeneutical deciphering can yield hidden insights about the self. In contrast to such a hermeneutics of the self for which psychoanalysis was the paradigm, Allport regarded confessional documents as stores of data for an analytics of the self. Today we can clearly recognize both forms of production of truth about the self. One can employ the classical procedure of poring over the textual details of a personal document to hermeneutically elicit its secrets. But one can also now process such documents—perhaps an autobiography, or maybe a collection of emails—with analytical tools that will run algorithmic analyses computing such data as frequency of word usage, syntactic patterns, and semantic profiles vis-à-vis other documents in a collection. These are the sorts of contemporary informatics of case materials that Allport anticipated.

Today such informational investigations seem straightforward enough. Some seventy-five years ago, though, a psychologist proposing a study of autobiography might have more likely deferred to traditional methods of interpretation. He or she might have undertaken what Allport described, discussing diaries, as "precious and meticulous exegesis."[216] But Allport seemed relatively uninterested in the work of the interpretation of meanings. He was rather focused on making use of documents as reservoirs of data. The difference is between documents that require a deeper interpretation and documents that bear on their surface information that can be processed: two styles for producing truth about the self.

Foucault famously notes in *The Will to Know* (i.e., *The History of Sexuality*, volume 1) that the "method of interpretation" central to psychoanalysis (and other hermeneutic models) did its work "by no longer making the confession a test, but rather a sign" such that it made sexuality into "something to be interpreted."[217] An earlier essay of Foucault's even ventures the vernacular of information theory to suggest that Freud "turned the verbal expressions of illness, hitherto regarded as noise, into something that would be treated as a message . . . a message of illness."[218] Foucault's statements concisely frame the difference separating the hermeneutic revelations of psychoanalysis from the analytics of personality informatics. The one is devoted to explicating "who one is" in terms of an essential deep kernel of selfhood that can be deciphered through signs that one cannot stop oneself from confessing. The other is in pursuit of a measure of "who one is" defined entirely in terms of observable metrics—for instance, the "confessions" solicited by a questionnaire that are taken not as signs for interpretation but as given data amenable to algorithmic analysis. If I am right that the difference is one of methodological technique, then this helps explain why Allport wanted to recruit the case study to his repertoire. What he objected to in Freudian analysis was its refusal to treat personal documents as sources of data that can be enrolled in algorithmic analytics on the same footing as data yielded by his personality questionnaires.

Critics today may object to information analysts like Allport—and all the progeny his program has spawned—that the work of interpretation is inevitable and as such cannot be so casually treated as a personality informatics would require. There are two halves to this objection. Concerning the first part (namely, that interpretation is inevitable), there is no doubt that Allport accepted the ineliminability of perspective that makes interpretation inevitable. In his visual arrangement of methods of psychology study canvassed in his *Personality*, all fifty-two methods surveyed are shown as radiating outward from a core of "observation and interpretation."[219] Interpretation pervades all scientific method. In *Personal Documents* too, Allport was clear on the point: "Preconceptions cannot be entirely avoided."[220] But considering the second part of the objection (namely, that interpretation requires our full seriousness), Allport just did not seem as impressed with the fact of the inevitability of interpretation as today's critical hermeneuts might expect. We might therefore pause with him. Is the pervasiveness of interpretation really something to be marveled at, or might it be true but banal? Just because something is inevitable does not mean that it is important that it is inevitable. Is it an option to acknowledge that interpretation pervades inquiry, forswear fundamentalist theories out of respect for this admission, and then casually move on? Or is the pervasiveness of interpretation something we must endlessly obsess over? Whatever one may think of these matters, Allport thought that the inevitability of interpretation could be quietly affirmed by psychologists so that they might then get on with their agendas.

What Allport himself moved on to was a conception of the case study in terms of informational analysis. He not only thereby enrolled the case study approach in psychological informatics, he also showed that it was possible for the psychologist to relax the energies often channeled to psychoanalytic hermeneutics and divert them instead to informational technologies. It is crucial to recognize that Allport well understood what was at stake here. His apparent positivism, whether we regard it today as felicitous or as pernicious, was anything but unstudied.

"A Considerable Army of Psychometrists"

Measurable personality was already solid by the time the great stock market collapse brought the Gatsbyesque exuberance of an era to a fairly screeching halt. Like the fascinating figure of Jay Gatsby, most Americans had been busily gaining personalities for themselves in the preceding decade. But whereas an enormous paper wealth dissipated, the paper personalities that had been manufactured by the analysis of questionnaires became ever more solid, reliable, and even necessary. In the same way that we can (and do) think of ourselves through Gatsby, we can understand one another (and ourselves) through personality traits.

By the 1930s, far more than just people had their personalities. Initially a way of conceiving of persons, "personality" had become so robust that it could be extended to anything. A host of books expressed the massive metaphoric extension. No less a household name than Emily Post wrote an influential interior-decorating book in 1930 titled *The Personality of a House*.[221] This followed on the heels of earlier authors ascribing personality to historical figures (Thoreau, Emerson, Napolean), social types (teachers, criminals), and a multitude of organizational forms (hospitals, cities). One commentator in 1930 could safely register that "personality is used to describe almost everything from the attributes of the soul to those of a new talcum powder."[222]

To the broader cultural obsession with personality, Allport had contributed a scientific understanding. The next wave of researchers were positioned to further leverage this contribution. Raymond Cattell—British born and brought to Columbia by Thorndike and then to Harvard by Allport—published his own book titled *Personality* in 1950.[223] Like Allport's earlier textbook of the same name, Cattell's was aimed at initiates to the field. But whereas Allport's needed to stabilize personality as a viable field of study, Cattell's could be hyperbolic about the field's importance: "Personality study is the natural hub upon which all more specialized sectors of psychology turn . . . the hub from which radiate all more specialized studies. . . . It is only by turning on this center that they make progress."[224] Four years earlier, Cattell had published *The Description and Measurement of Personality*, the first of a spate of contributions for advanced researchers in a now-stable field. At the end of his first chapter, Cattell confidently invoked the field's authority in noting "the competence of a considerable army of practicing psychometrists."[225] This was an army of data recorders and information analysts whose chief armaments were the questionnaire and the algorithm, each soldier drilled in the use of instruments that were less than three decades old.

As psychological science expanded its authority in the decades following World War II, more and more came under its expert survey.[226] One politicized domain where it accrued increasing prestige was that of race. Allport published his landmark *The Nature of Prejudice* in 1954. It is a book that is still cited to this day—for instance, by political theorists wrestling with what to do about the injustices of racial segregation perpetrated by the American government in the very years leading up to Allport's study.[227] But before a psychologist like Allport could run studies showing the tonic influence of cross-racial contact on undoing prejudice—and thus also before political theorists could cite those studies as justification for particular approaches to rectifying racial inequalities—there needed to be first a way of making race fit the formats of the data systems such studies depend on. Racial identity, just like psychological identity, had to be made to become data.

Segregating Data:
The Informatics of Racialized Credit, 1923–1937

Redlining Race

Ta-Nehisi Coates's 2014 article in the *Atlantic* on the seemingly modest topic of the history of the American home loan industry was nothing less than explosive.[1] The article brought a new wave of attention to the history of redlining, a euphemism that refers to the practice of denying home loan applications based on the racial characteristics of a property's neighborhood. Assessors, lenders, and underwriters in the 1930s and 1940s had employed government-produced rating maps on which were drawn red lines around areas deemed too risky for loans. Areas were often rated riskiest because of the presence of nonwhite (almost always African American) residents. Those red lines on paper maps soon became high walls in America's cities.

Coates's article opens with the story of Clyde Ross, a Mississippi-born transplant to Chicago, one of six million black Americans who moved northward during the Great Migrations of the early to mid-twentieth century. Ross left southern segregation and lynching only to find himself landed squarely in the subtler and slower violence of northern racisms. What Ross sought was the American dream: a home for his family. But he bought his home on contract—a midcentury credit instrument that left (mostly black) buyers without equity and legally hamstrung if they missed a single payment.[2] Ross nearly lost his family home, but he managed to hold on by taking three jobs (at Campbell's Soup, at the post office, and delivering pizza). He was one of the "exceptional ones," notes Coates, because for every Clyde Ross, "there are so many thousands gone."[3] When Coates met him, Ross was ninety-one years old and still living in the Chicago home he had barely managed to buy on contract. "You could fall through the cracks easy fighting these white people," he told Coates.[4] And since many thousands did, Ross's home was by 2014 massively depreciated. His North Lawndale neighborhood bears the telltale signs of decades of redlining, contract selling, and racial covenants: a 92 percent black neighborhood with a 43 percent poverty rate that doubles Chicago's average, a homicide rate triple that of the city, and an infant mortality rate twice the national average.[5]

Coates's unambiguous argument is that government redlining and the contact selling it made possible are injustices that enjoin an obligation to consider the question of racial reparations: "The crime indicts the American people themselves, at every level, and in nearly every configuration."[6] Coates's title, "The Case for Reparations," forcefully states the point.

In assembling his argument, Coates excavates the devilish details of redlining maps as used at the great home finance agencies of the Depression: the Federal Housing Administration (FHA) and the Home Owners' Loan Corporation (HOLC). The social scale of those maps is dizzying in light of how far the FHA reached into the lives of most Americans. Historian Kenneth Jackson, whose archival research in the 1980s uncovered the extent of government redlining, noted that "no agency of the United States government has had a more pervasive and powerful impact on the American people over the past half-century than the Federal Housing Administration."[7] If you are white and your family are not recent immigrants, it is extremely likely that your forebears directly benefited from the FHA subsidies denied to black families—even if your forebears were hardworking, bootstrap-pulling white Americans, they (and in turn, you) were beneficiaries of a handout not extended to hardworking bootstrap-pulling black Americans. The racialized effects of redlining thus persist some eighty-plus years later, as can be gleaned from a recent web project assembled by the National Community Reinvestment Coalition that overlays the geographies of the HOLC maps with current demographic data on residential racial segregation and socioeconomic status.[8] There you can find a map of Clyde Ross's neighborhood circumscribed by those historic red lines squaring up exactly with geoinformatics data showing its "low-to-moderate income" status today.

The Informatics of Race

As is the case for any injustice, redressing redlining will require specific policies implementing specific mechanisms toward specific goals. This is true for whatever political vision of racial justice we pursue, be it reparations, integration, or desegregation, for any such vision will require implementation through particular mechanics that will countervail the entrenched mechanisms of this injustice.[9] So what were the mechanics through which redlining was implemented? How was it operationalized? On what techniques did it rely? These questions are central to any normative reckoning with redlining. What form would a more perfected justice need to take such that it would not leave intact the technological facilities of redlining?

I argue in this chapter that redlining depended on what I call an "informatics of race" that reformatted long-standing racial differences as distinctions in data. The specific techniques funding government redlining from 1934 onward included information technologies put into motion in the real estate industry beginning in 1923. These industry-specific technologies were part of a wide network of racial datafication enacted in that decade. After the installation of this apparatus, race had crossed a threshold that we continue to live on the other side of.

Consider James Weldon Johnson's 1912 autobiographical novel *The Autobiography of an Ex-colored Man*.[10] Johnson's account of passing as white would have been legible to his first readers in the 1910s, but would after that decade become much more controversial, and perhaps for that reason much more popular (the book first gained wide readership when republished in 1927). At that point, passing could no longer be just a matter of declaring oneself differently to a new community. It would henceforth be enrolled in data technologies that would always announce our race for us in advance of any arrival. Interrogating the formation of these data technologies offers a way of seeing how sometimes-transformable race was made into ever-obdurate data.[11]

But what is at stake in attending to the technical aspects of race and more specifically to its particular informatics techniques? Why engage in a discussion of the history of race in terms of racializing information technologies?

Answering these questions turns on a conception of the connection between normative and diagnostic theory. A diagnostic analysis of how technologies of politicization work is central to any project of normatively assessing or reconstructing the practices employing those technologies. Normative assessment and reconstruction both depend on understanding how injustices have been perpetrated, propped up, and passed along. For if we fail to understand the specific mechanics by which injustices are operationalized, then we may end up leaving those injustices in a critical vacuum where they are likely to remain intact. Thus the normative work of determining where we have an obligation to rectify injustices requires historical diagnostic arguments detailing how injustices have emerged, how they operate, and how they are maintained today. It thus requires, among other things, attending to how specific techniques operate.

Closer to the case at hand, my specific argument is that assessing and reconstructing practices of redlining requires (in addition to much else) understanding the data technologies that allowed for redlining in the first place. I thus aim to fill a void of critical technical analysis in the recent voluminous scholarship on state-sponsored real estate segregation in the 1930s on, for no existing studies that I am aware of examine the development of appraisal techniques by private industry in the 1920s. Yet these very techniques were later leveraged into government redlining in the 1930s. Filling this void will be a diagnostic contribution to recent work in the normative political theory of race. But, more central for my broader argument, this will underscore the value of a critical interrogation of informatics techniques.

With respect to the political theory of race, the exclusive focus on state-sponsored activities in this literature certainly makes sense in the context of arguments that we have an obligation to remedy these injustices.[12] And yet an exclusive focus on state institutions, laws, and policies leaves unexamined those techniques of racial informatics that easily shuttle between state and nonstate actors. I argue that an account of how racializing technologies work needs be part of normative argumentation about racial injustice.

I propose this expansion with the idea that it is already motivated by the methodological perspective of "realist" or "nonideal" theory that informs much of the best work in the political theory of racial inequality. Paradigmatic of this perspective is work by Elizabeth Anderson, whose book on racial integration has been called "the most significant and comprehensive scholarly work calling for racial integration as an imperative of justice."[13] Anderson argues that seriously addressing racial injustices requires a philosophical methodology of "nonideal theory" that "integrates research in the social sciences in ways not ordinarily found in works of political philosophy."[14] Anderson's approach to nonideal or realist political theory is, like mine, funded by methodological elements from John Dewey's philosophical pragmatism.[15] But whereas Anderson puts pragmatism into motion as a normative social theory alongside contributions from sociology and social psychology,[16] my approach plays up pragmatism's emphasis on historicity and historical forms of inquiry, a combination also motivated by philosophical genealogy.

Genealogy is explicit about the need for attention to histories of political practice in light of their technical minutiae. Foucault once described his analytics of power as an inquiry into "the techniques and procedures by which one sets about conducting the conduct of others."[17] Latour, later citing Foucault, stated the idea even more strongly in claiming that "the only way to understand how power is locally exerted is thus to take into account everything that has been put to one side—that is, essentially, techniques."[18] Such a perspective helps us remain resolute in our attention to political technologies despite their status within the canon of modern critical analysis as negligible elements of political mobilization. Following these pathways into the techniques of racializing informatics in the early twentieth century, my perspective resonates with recent work that examines processes of imbricated technologization and racialization. For example, my focus on racializing technologies builds on Geoffrey Bowker and Susan Leigh Star's Foucault-and-Latour-inspired analysis of "the actual techniques used to classify people by race" in apartheid South Africa.[19] But more poignant than now-classical contributions to science and technology studies is recent work in critical race studies drawing on the critical theory of techniques.

One such lead for my analysis is Debra Thompson's argument in *The Schematic State* that census categories are "an evolving race-making instrument."[20] The census is of course a biopolitical instrument par excellence. But Thompson's study helps us see that there is also a politics to census categorization and classification.[21] What is striking about her analysis for my argument is that, despite a rich history of biopolitical uses of census data across the nineteenth century, the infopolitical formatting of racial census categories became particularly charged in the early twentieth century and did not stabilize until after the 1930 census cycle, when those categories entered a period of quiet that persisted for almost half a century in advance of today's renewed census debates.[22]

Perhaps the central exemplar for my approach is Simone Browne's study of the surveillance of blackness in *Dark Matters*. Browne's book assembles a series of historical and contemporary technologies that together form an apparatus of "racializing surveillance."[23] In so doing, she looks through—but also beyond—Foucault's analysis of surveillance technology as implementing disciplinary power.[24] Browne conscripts a counterhistory of racializing surveillance techniques ranging from brutal technologies of the branding of flesh to paper machines for building racialized databases of travelers. Browne's assembly shows how "racializing surveillance is a technology of social control where surveillance practices, policies, and performances concern the production of norms pertaining to race . . . [and] reify boundaries, borders, and bodies along racial lines, and where the outcome is often discriminatory treatment of those who are negatively racialized by such surveillance."[25] This approach exemplifies a critical strategy of interrogating technologies for their normative accumulations. As Browne writes of another example elsewhere, "Profiling technologies can intersect with existing social hierarchies and inequities to perpetuate stereotypical representations, racism, and criminalization."[26]

To these and other analyses of racializing technologies, this chapter adds a detailed inventory of a technology of racialization rooted in information. The specific focus is on what I conceptualize as "the informatics of race," which I also sometimes refer to as "racializing informatics." I employ these terms as referring to a dimension of racialization. Racialization is "the extension of racial meaning to a previously racially unclassified relationship, social practice or group."[27] My claim is that racialization in some instances depends on the work of the informationalization of race—that is, the extension of data to previously dataless concepts of race. The informatics of race can be used to sustain and strengthen subtle aspects of structural racism that I conceptualize as "technological racism." I employ this label to refer to a little-discussed dimension of structural-institutional racism in contrast to other of its more familiar dimensions such as racist policies, procedures, organizational habits, and social customs.[28] The notion of technological racism thus draws attention to the technical mechanics of how purportedly neutral information systems can unexpectedly embed racial bias, discrimination, and inequality. Recent work on algorithmic bias in automated intelligence machines leveraging massive databases exemplifies the need for an analytical category capable of bringing into focus technological racism in its contemporary and, I add, entrenched historical forms.[29]

In employing the concepts of "the informatics of race" and "technological racism," it is crucial for my argument that I maintain a distinction between the two. It is not my argument that the informatics of race are in every instance instruments of racism, nor is it my argument that the informatics of race can ever be neutral. My claim is that the informatics of race admits of multiple valences of politicization. Racializing data can be leveraged not only for racist technological infrastructure like redlining maps, but it can also be used within technologies countering racism—for instance, in uses of racial census data to demonstrate discriminatory disparate impact.[30] Keeping this distinction in view, I argue that racial data has become one vector of power through which racism maintains itself, and that this maintenance is not therefore inevitable in any context in which the deployment of racializing data is unavoidable. It is because of that non-inevitability that it can be valuable to study the specific mechanics of the informatics of race.

What is not inevitable may nonetheless be actual. The history here under survey is one in which racial data was time and again leveraged into technological racisms. To get a sense of why this matters, imagine an alternative history of the twentieth century in which connections between data and race had not been forged. No redlining, no credit inequity, no financial racism, and no prejudicial inequalities on any scale large enough to require statistical reasoning. In such an alternative history, racism would surely have sought, and may indeed have found, other means of maintaining itself. But those alternatives would have produced a racism dramatically different from that which persists today. This illuminates how deeply today's practices of racism depend on data. This is a specific dependency rooted in a specific history. Failing to attend to such specificities hobbles our efforts in normative assessment and reconstruction.

In the context of the book's broader analysis, my argument about the politics of racial information would seem at first blush to be one of the easiest arguments I could make on behalf of my thesis that information has become political—for we are so accustomed today to understanding race as ineluctably political. Yet in fact this argument is one of the most difficult that I navigate in this book. This is because the way in which we think of anything that is racialized as thereby political overdetermines from the outset my argument that the data dimensions of racial data are political. My claim is that racial data are doubly political—they are political insofar as race is ineluctably political (call this the "politics of race" dimension), but they are also specifically political with respect to the data that is there at play (call this the "politics of data" dimension).

My argument, then, is that the history of the emergence of the informatics of race is a partial history of an underexamined vector of infopower that too much pervades our present. This historical perspective is needful just insofar as, in the words of James Baldwin, "the past is all that makes the present coherent, and . . . the past will remain horrible for exactly as long as we refuse to assess it honestly."[31] Some readers may find it a stretch to apply the word "horrible" to something as austere as an informatics apparatus. But consider: before the bloody rivers of violence of the Jim Crow era could be dammed, white supremacists in those waning years of the lynch mob had to first convince themselves that they could shift supremacist strategies to efforts that have been retroactively referred to as "statistical racism" and "statistical ghettos."[32] We need to interrogate these seemingly plain histories of data that bear within them a more dangerous past. We need to focus attention on the ways in which whiteness and blackness have become innocently representable in little boxes that we unthinkingly tick and check (a mechanics also at play in other racial and ethnic categories whose histories I do not interrogate in this chapter, and not for their unimportance, but rather because of their more limited place in the specific history of redlining). Such information technologies transformed prior acts of racial categorization into informational inputs prepared for processing and redistribution. This was the crucial moment of the datafication of race. In order to make racist uses of racial data, race has to first be made to become data, which is to say that racial identity has to be made to depend at least in part on data.

A Genealogy of Redlining: An Emergent Informatics
of Race, ca. 1923–1937

In 1951 the American Institute of Real Estate Appraisers (AIREA) published a half-manual-half-textbook titled *The Appraisal of Real Estate*. The handful of racist assumptions plainly stated in the book typifies views of race in real estate valuation at midcentury.[33] The professionalized position from which these statements were published was formidable. AIREA was established in 1932 as a branch of the National Association of Real Estate Boards (NAREB).[34] One commentator later described the relation between NAREB and its AIREA subsidiary as follows: "The appraisal group is the intellectual wing of the NAREB. . . . They are the social and intellectual guides in matters of real estate in which official and unofficial agencies are involved."[35]

Nearly twenty years after its founding, the education committee of this intellectual wing of NAREB asserted in its half-manual-half-textbook that, "when a new class of people of different race, color, nationality, and culture moves into a neighborhood, there is a tendency on the part of the old inhabitants to think that the neighborhood is losing desirability; and thereafter they tend to move to other districts."[36] This statement is the first of an enumerated list of "factors that contribute to the destruction of value." Its terse style is typical of its professionalized origin. It surrounds itself in the aura of fact. It quietly asserts. It does not condescend to defend. It does not worry over muddled details of moral validity or epistemic justification. It merely observes what would be there for anyone who bothered to look.

Such a statement would not have been possible without a whole flotilla of efforts that had been in motion since the mid-1920s. Three decades into their development, these efforts could be stated in the confident prose of a plain-spoken vernacular appropriate to the format of a technical manual for an audience of trained technicians who saw themselves as embodying professional expertise in a field whose work was newly in possession of scientific merit. In these contexts, race could be presented in the raw material of categories and numbers since professional real estate appraisal had already performed the work of rendering race in terms of data and stabilizing racial information as material for work. This was no small matter. In three short decades, racial identity had been coded into data that could be—and indeed, still is today—taken as an innocent mass on which algorithmic ingenuity can be unleashed.

Like much that grows to powerful political proportion, redlining had tiny beginnings, first taking shape through a handful of real estate men from Cleveland and Chicago before eventually disseminating nationally through the work of a not-much-larger number of men at NAREB and AIREA who sought to give to real estate brokerage the imprimatur of professionalism and scientificity. Foucault's work describes disciplinary power as emerging through "humble modalities" and "minor procedures" rather than the majestic forms typical of so much political theory.[37] If real estate appraisal techniques seem insignificant in comparison to the full magnitude of American racism, then this chapter is a reminder of how even assemblies as gargantuan as racism have to be cobbled together out of minutiae like standardized forms and the formats imprinted on those forms. Such microdimensions of power often prove difficult to resist precisely because we tend not to see power operating at those tiny end points where it often has its greatest hold.

Techniques of appraisal that emerged in the 1920s and consolidated in the 1930s offered a way of maintaining racialization within contexts where many sought new means of maintaining racial hierarchy as a response to changing landscapes of race. This was a moment in which faiths in white supremacy found themselves facing new threats. One was the declining status of scientific racism, which by the 1920s was being subject to a severe scrutiny that would soon bring that program to a more or less definitive end. Another was increasing public resentment about the unpredictable racisms of Klan violence and southern lynching that reached its high point in the 1890s and began to decline quite sharply in the early 1920s. As the biopolitics of scientific racism and the sovereign violence of lynch mobs lost their standing, white supremacy went searching (not necessarily self-consciously) for new means of maintaining itself. What resulted was a collision between novel mechanisms and entrenched but disrupted habits.

In creating a new kind of space for racialization within real estate appraisal, there were two significant moments. The first moment, running from 1923 to 1925, was one of first formulations. A range of precipitating factors, including aspirations to professional and scientific status, opened opportunities for the introduction of new forms of racialized evidence in the context of improving valuation methods. These years were witness to the earliest formulations of race-conscious residential property valuations in a range of treatises, textbooks, and articles in professional journals. As racialized appraisal practices matured, a second moment, from 1931 to 1937, was one of consolidation and systematization. These years were witness to standardizations that could be performed with prefabricated equipment geared to appraisal professionals: standard printed blank forms, standard filing systems, and technical manuals specifying how to use these technologies to maximal advantage. During this second moment there was also a redeployment of these consolidated techniques of racialized accounting by the state agencies soon to be responsible for redlining. In a few short years—from inception in 1934 to the completion of technical programs of appraisal in 1935 and demographic survey in 1938—the state built technologies of racism into its most important welfare program in a way that went largely unnoticed because it relied on purportedly neutral racial data.

The aftermath of these techniques has persisted for decades. Many have recounted the tragic histories of America's racialized unfreedoms: postwar projects including the production of the sprawling suburbia of the Levittowns, real estate contract selling in cities like Chicago, and the self-fulfilling prophecies of racialized credit.[38] A neglected aspect of this postwar history is the role of the informatics of race.

PRECIPITATING AN INFORMATICS OF RACE, 1910–1924

I begin with a brief consideration of some motivating problematics in the real estate industry within which the informatics of race would gain currency. The most important background feature is that of residential segregation. According to the 1993 study *American Apartheid* by sociologists Douglas Massey and Nancy Denton, black-white urban segregation in 1940 was at 81.0 percent on a standard index of "dissimilarity" (a representation of the percentage of minorities who would have to move to achieve a residential pattern in which every neighborhood replicated the racial composition of the city on the whole), but in 1910 had been at just 38.8 percent.[39] At the outbreak of the Civil War, black-white segregation stood at 29.0 percent. Massey and Denton conclude that, "before 1900, blacks were not particularly segregated from whites" and that there is "little evidence of ghettoization among southern blacks prior to 1900."[40] This claim is startling to those accustomed to progressivist accounts of increasing integration over time as a legacy of the Civil War abolition of slavery. Yet the racial segregation of urban America did not precede the Civil War; it followed it by about half a century. Segregation in the north was at its highest peak around 1950, when the segregation index of many cities (including Chicago, Cincinnati, Cleveland, Indianapolis, St. Louis, and some southern cities like Atlanta and Houston) climbed above 90 percent.[41]

Residential segregation began its increase in the 1910s as northern whites confronted half a million African Americans moving northward as part of the Great Migration.[42] Many communities, particularly in border and Midwestern states, sought insulation from the influx. The comparative integration that had previously characterized these communities devolved into explicit, and often explicitly racist, segregation.

The first technique to bear the weight of segregation was the zoning law. A Baltimore ordinance enacted in December 1910 is, according to Charles

Abrams in 1955, "generally referred to as the first racial zoning law in America."[43] But soon the Supreme Court, in its 1917 *Buchanan v. Warley* decision, invalidated these ordinances.[44] The unanimous ruling was that these ordinances were not legally enforceable on pain of violating equal rights to the disposal of property. In other words, residential segregation could not be secured by mechanisms of discriminatory law.[45]

Yet *Buchanan v. Warley* forbade only the use of the public institution of law to secure discriminatory segregation. It did not rule on the legality of private acts of discrimination. Thus a wide path remained open for the development and deployment of alternative mechanisms of segregation. Before turning to the most lasting mechanisms developed in response to this matter, I want to briefly canvass two that made a large impact even though they were eventually halted by another Supreme Court decision in 1948.

One such mechanism was the restrictive covenant—a contractual agreement among property owners belonging to a particular neighborhood improvement association that forbade minorities from owning, and sometimes occupying and leasing, property subject to the covenant. In a 1969 book, Rose Helper traced the origin and spread of these covenant.[46] She speculated that the first covenant was innovated in 1914 by a Harlem real estate agent who proposed in the *Real Estate Bulletin* (published by the Real Estate Board of New York) to "arrange with the property owners to rent their properties to white tenants."[47] Soon after, in 1917, the Chicago Real Estate Board convened a committee to explore "the invasion of white residence districts by the negroes"—among its recommendations was "a propaganda through its individual members to recommend owners societies in every white block for the purpose of mutual defense."[48] Military metaphors of "invasion" and "infiltration" would soon figure prominently in how professionalizing real estate agents discussed race. But so too would their desire to distance themselves from attitudinal racism. The Chicago committee made clear that it was concerned "with a financial business proposition and not with racial prejudice," even going so far as soliciting "the co-operation of the influential colored citizens."[49] The covenant technique, which rapidly spread in the 1920s,[50] was a first promise of an alternative pursuit of racism that would claim to be both scientifically valid and politically neutral. Any technique that could claim both of these would, in that historical context, have much going for it.

At the same time, the real estate men at NAREB sought a second mechanism for racial segregation that could also be seen as morally unobjectionable and economically favorable to all. This was NAREB's "Code of Ethics." NAREB adopted its first such code in 1913.[51] A decade later, a revision was moved at the association's 1924 annual meeting and was adopted by the

2,692 member-attendees from thirty-nine states.[52] Among the thirty-five articles of the code they endorsed, the penultimate rule, article 34 of part 3 ("Relations to Customers and the Public"), enjoined a strictly economic duty to concern oneself with the racial composition of neighborhoods: "A Realtor should never be instrumental in introducing into a neighborhood a character of property or occupancy, members of any race or nationality, or any individuals whose presence will clearly be detrimental to property values in that neighborhood."[53] While covenants and ordinances had employed a morally charged language of racial exclusion, NAREB's article 34 attempted to solidify segregation as a pure function of "property values." Though economics can never fully avoid ethics, the point is that there was a polished veneer of professionalism through which NAREB sought to sustain differential treatment. The technical apparatus of a code of ethics would prove convincing to many in its promise of being at once epistemically warrantable and politically neutral.

Yet in 1948 such codes of ethics and neighborhood covenants met the same fate as their municipal zoning cousins.[54] In the work they performed over two decades they indexed racism's migration from the brutality of the noose to the refined technicity of such matters as accounting practice. They helped set the stage for residential real estate appraisal as a procedure whose informatics would help stabilize the place of racialization in the context of property values.

PROFESSIONALISM AND SCIENTIFICITY IN APPRAISAL, 1923–1925

Alongside the covenants and codes whose expiry would be forced in 1948, there emerged another technology that would eventually consolidate into a form that appeared (at least, for a while longer) politically neutral. This technology claimed its neutrality as a function of its professional and scientific competencies with data. Through the sheen of professionalism and scientificity, a surprisingly small handful of reformers were able to bring an informatics of race to bear on real estate appraisal.

In the 1920s, the real estate industry was in a fever pitch of professionalization. Sinclair Lewis's 1922 novel *Babbitt* presented a clear-eyed portrait of the personality of the real estate man in its title character. Here is Babbitt trying out on a fellow Realtor a speech to be delivered to his State Association of Real Estate Boards: "What is it distinguishes a profession from a mere trade, business, or occupation? What is it? Why, it's the public service and the skill, the trained skill, and the knowledge and, uh, all that, whereas a fellow that merely goes out for the jack, he never considers the—public service and trained skill and so on."[55] Lewis may have been mocking the Babbits of his day and their corking "specifications of the Standardized American Citizen," but the truth of the caricature is in Babbitt's unstudied awareness of the need to distinguish himself by claiming for his vocation a special set of skills.[56]

No small part of this professionalization was the reconstitution of core real estate practices as scientific, objective, and descriptive. The practice of appraisal was professionalized by bringing standardization into view as a possible goal. Such standardization was established through the work of appraisers emphasizing the scientificity of their work. When charged categories like race would appear in real estate accounting, these categories could accordingly be construed as purely descriptive.

NAREB was founded in 1908 (under its first name of National Association of Real Estate Exchanges that lasted until 1916) and soon became an effective national organ for professionalization.[57] In the late 1910s and early 1920s, journals like *National Real Estate Journal, Annals of Real Estate Practice*, and *Real Estate Record and Builders' Guide* began to feature articles describing professional and scientific programs for industry practice. One key figure in this visioning was Herbert U. Nelson, a longtime correspondent for, and eventually editor of, the *National Real Estate Journal (NREJ)*. In a 1923 article titled "Systematizing Real Estate Education," Nelson adumbrated a two-year course of study that would lead to a thorough training in the technicalities of the real estate profession.[58] At the center of his vision—and recommended as the first text to be digested by his would-be professionals, was the still-unpublished *Principles of Real Estate Practice* by Ernest M. Fisher, assistant executive secretary of NAREB (the executive secretary to whom he was assistant was Nelson).[59]

It is striking how small was the circle of men who instigated the push for real estate professionalization and how successful was that push, given their number. One way to get a grip on the size of the group is to note its geographical centers: almost all of them were from either Chicago (where NAREB was headquartered) or Cleveland, with a few additional early advocates coming from other cities in Illinois and Ohio. Professionalism soon spread nationally. But the movement began at a handful of desks in the Midwest. As in so many matters, a few people who were relatively powerless and previously unconnected came to articulate a vision that would soon enroll an entire nation of homes and their owners.

Nelson could recommend Fisher two months before his book was published because it had already been decided by NAREB that Fisher's book would be the introductory text for a course of professional education. As

Fisher recounts in his prefatory notes, NAREB sponsored a conference in Madison, Wisconsin, early in 1923 gathering members of NAREB, the United YMCA Schools, and the Institute for Research in Land Economics and Public Utilities.[60] The latter was the research arm of influential University of Wisconsin economist Richard T. Ely, who wrote a preface to the volume and published it in his Land Economics series. The 1923 conference attendees "outlined and adopted a course in real estate practices and principles that could be used in educational institutions as a standard."[61] As part of that course, "the outline of [the] book was approved by the conference as suitable for the first elementary course designed to introduce the student to the whole field of real estate practice."[62] Thus was it proclaimed in the book's front matter, opposite the title page, that the title was "Volume I in the Standard Course in Real Estate Outlined by the Joint Commission" established by the three organizations who participated in the conference.[63] The book was published in October 1923. At the end of the year, Fisher published an article in the *NREJ* laying out a fuller course of professionalized study that slightly revised Nelson's proposal.[64]

Less than two years later, there appeared in the pages of the *NREJ* an article that exemplifies how much the idea of a professionalized course of study had taken hold. In the August 1925 issue, Henry Grant Atkinson of Cincinnati published his plea for a "Standard Test for Real Estate."[65] Atkinson began by praising the US Army's mass-scale intelligence testing during World War I.[66] He then excitedly reported on such a "standard test" recently devised "for persons entering or intending to enter the real estate business."[67] He even went so far as to claim that, "so far as the writer is aware, the NAREB is the first professional organization to develop a standard education test for the use of its members and the schools serving it."[68] It is unlikely that this claim was true; but that matters little, for what is of interest is simply that it was asserted. What that assertion expresses is a proud professionalization in complete consciousness of itself. Atkinson's argument was not an isolated incident. A similar article by Thomas Nelson titled "A Standard Test in Real Estate" appeared that same year in NAREB's *Annals of Real Estate Practice*.[69]

In that same year, again in the *Annals of Real Estate Practice*, there appeared an article on the subject of a real estate training course by Fisher. The article included a subsection on appraisal in which the following texts were recommended: Frederick Morris Babcock's 1924 *The Appraisal of Real Estate*, John Zangerle's 1924 *Principles of Real Estate Appraising*, a 1920 reprint of an older 1903 text titled *Principles of City Land Value* by Richard M. Hurd, and Stanley McMichael and Robert Bingham's 1923 *City Growth and Values*.[70] Babcock, from the Chicago suburb Evanston, had his book published under the imprimatur of NAREB in the same Ely-edited series as Fisher's text, and was it presented as volume 3 in the "Standard Course in Real Estate."[71] Zangerle and McMichael, both from Cleveland, published their texts with the Stanley McMichael Publishing Company. These same few texts were cited over and over in NAREB publications.[72]

All these texts expressed strong aspirations to scientificity and professionalism. Those aspirations had been ably summarized by Frederic Howe in his 1910 address, "The Scientific Appraisal of Real Estate."[73] Howe defended a particular system of appraisal, the Somers system, whose method he thought "scientific" because, rather implausibly, it involved "a law of causation."[74] But what Howe really sought—and what would be increasingly achieved a decade later—was a reliable algorithm for appraisal. Therein lay the path of a data science, and through that path would arrive the status of being a professional. That aspirational path is explicit in the appraisal texts recommended in NAREB's publications.

The 1924 Zangerle book recommended by Fisher opens with a chapter on "Scientific Appraising": "To secure results, there should be substituted system instead of caprice, standards in place of guesses."[75] Fisher in his own 1923 book observed that the aims of NAREB included bringing about "more scientific, uniform, and efficient methods of dealing in real estate."[76] He also referred to science in the context of discussing appraisal: "Justice to all parties interested in the appraisal requires that it be made with all skill and science available."[77] Fisher's references to professionalism were even more effusive. The subject not only received an entire chapter but was also central in the book's first chapter, where real estate dealers were analogized to lawyers, doctors, and priests in the context of an argument laying down as the key to professionalization the appearance of "textbooks" as vehicles for extending a student's "information."[78] Similarly, the 1923 book by McMichael and Bingham (not yet boosters of scientificity) emphasized professionalism throughout: "Land valuation is not an exact science—it is a profession excelled in only after years of intensive study and wide experience."[79]

Despite some initial hesitation, all these authors soon viewed scientificity as integral to real estate professionalism. By the time the widely circulated *McMichael's Appraising Manual* was published in 1931, its author had converted to science: "Real estate appraising is definitely being standardized through the experience and practice of men and organizations in all parts of the United States and Canada who encounter the same class of valuation problems and who, by various methods and processes, succeed in solving them in a scientific way."[80] Fisher in his 1930 *Advanced Principles of Real Estate Practice* observed that "improved technique is substituting scientific for haphazard procedure and creating greater efficiency and reliability in the practice of the vocation."[81] Fisher's idea of improved technique was basically that of data standardization, or "the increase in technical information available, accompanied by a tendency toward greater standardization of practice."[82]

The most interesting of these early texts is Babcock's 1924 *The Appraisal of Real Estate*. The first sentence of its preface clearly captured the tendencies of the time: "Real estate appraising is now looked upon as scientific and professional in its character."[83] This was an overstatement, but the important thing in such cases is not so much the truth of what is stated as the statement's expression of a certain aspiration to become true.

Babcock's work is, fascinatingly for 1924, presented in the technical vernacular of nascent information science. In chapter 4, "Appraisal Procedure and Methods," Babcock offers an analytical taxonomy of "the program of procedure in the making of an appraisal."[84] The overall "program" is presented as divided into three "subprograms": the Data Program, the Computation Program, and the Report Program.[85] Each subprogram is presented in terms of the way in which it handles information: information inputs, information analysis, and information outputs. The Data Program involves "selection of sources of information"—collection, testing, investigation, research.[86] The important thing here is to "rely on such information a great deal more" than on "unsupported judgment."[87] The Computation Program involves a series of methods, each tailored to the specifics of the property being valued. The methods are explicated as procedures that the appraiser should carefully follow in every detail. Lastly, the Report Program is a selection procedure for both determining "kinds of information deliberately *left out*" of final appraisal reports and ascertaining the "specific form" for data included in them.[88]

There are two notable features in Babcock's presentation. First is his usage of a vocabulary of informatics that at that time had still not congealed around metaphors now habitual for us today.[89] Today we want to think of programs as defined chunks of code that "run" inside of the many kinds of computers increasingly populating our lives; Babcock uses the term *program* in a less metaphorical sense to refer simply to ordered procedural operations. This suggests a second notable feature of his presentation: the chapter impresses on the contemporary reader the image of a perfect metaphor for a computer, a machine, an algorithm. Like any algorithm, Babcock's is robust—it simplifies, to be sure, but it does not obviously oversimplify. Babcock is explicit that each of the subprograms admits of a multiplicity of methods of operationalization according to the dictates of the situation at hand.[90]

The statements of professionalization and scientization just recounted all produced a promise of political neutrality. I do not aim to dispute this polished faith, but rather to examine just what it was used for. In the work of the small number of Midwestern men who put these tendencies into motion, we also find a series of plainspoken observations about the role of race in real estate valuation. Through the vectors of professionalization and scientization the NAREB textbooks leveraged a whole ensemble of techniques of racialization.

The texts that Nelson and Fisher praised—which were also the texts that would have the greatest impact on the profession—frequently made race an explicit part of the appraisal equation, thereby sedimenting an algorithmic technique of factoring race into price.[91] Exemplary of this are the three texts just noted for their emphasis on professionalism: Fisher's chapter on appraisal from his October 1923 textbook on real estate practice, McMichael and Bingham's November 1923 study of city values, and Babcock's September 1924 technical treatise on appraisal.[92]

McMichael and Bingham's chapter 17 was titled "Foreign and Racial Settlements." The chapter's first sentence doubled down on its titular equivocation: "While settlements of foreign born residents or colored people sometimes have the effect of making cities grow rapidly it is significant that in some instances, notably where negroes congregate, land values are depressed."[93] What is so striking about this language is not its presence in a 1923 book, but rather that its authors offer no justification for such claims about racial difference. They later asserted that "property values have been sadly depreciated by having a single colored family settle down on a street occupied exclusively by white residents."[94] The claim is empirical, but no evidence was given for it, not even an anecdote. McMichael and Bingham time and again relied on such empirical claims, but in so doing they appear not to have wanted to condescend to offering evidence for those claims. Some readers noticing the lack of proof might have concluded that such claims were untrue, or at least unproven. Others might have thought that these claims were true, but unproven, and so in need of proof. A scientific and technical proof for the relation between race and value is precisely what the appraisal profession would begin to try to assemble.

Fisher in his 1923 text took a tack quite similar to McMichael and Bingham's, but he went one small step further in realizing that some work had to be done to organize his more empirical-sounding bids. In calm and confident prose, Fisher noted that "it is a matter of common observation that the purchase of property by certain racial types is very likely to diminish the value of other property in the section."[95] Still without offering evidence or anecdote, Fisher's discussion did, however, add context for this claim in the form of a taxonomy of appraisal that specifies where it applies. Fisher's chapter 7, "The Valuation of Real Estate," was divided into two sections, one on farmland and one on urban real estate; the latter was further subdivided into manufacturing, commercial, and residential real estate.[96] Race was mentioned as a factor only with respect to residential values. As appraisal informatics were later leveraged to solidify relations between race and value, this distinction persisted and the racialization of value was generally applied only to the case of residential property.[97]

Babcock in 1924 opened his discussion in chapter 3, titled "The Elements of Real Estate Value," with the observation that among the "factors which have of necessity been influential in determining the locations of cities" are climate, intellectual considerations, and "racial congeniality."[98] This last factor Babcock explicated in a section titled "The Effect of People on Values" in which it was asserted three different times on a single page that race is a factor: "Residential values are affected by racial and religious factors. . . . And so the habits, the character, the race, the movements, and the very moods of people are the ultimate factors of real estate value. . . . The real factors are buying mood, hours, purchasing power, motives at the moment, directions of movement, race, occupations, religion and standards of living."[99] Thus would race need to be factored into the scientificity of the tripartite appraisal algorithm discussed above. Babcock was clear that his Data Program for obtaining "all manner of data concerning the neighborhood and its future" includes information on "racial movements."[100]

Babcock's reference to racial movement foreshadowed a future development in residential appraisal. In ascertaining the racial composition of what Babcock called "districts," the appraiser can collect data that conveys a frozen portrait or they can seek data that shows directionality. Only the latter will be of use in forecasting in a "dynamic society" in which "the characters of the districts are varying and in which the population is increasing or diminishing."[101] This attention to neighborhood change would come to be characteristic of residential appraisal techniques. Not only would later appraisers express explicit concern for race "infiltration" and "invasion" but they would also seek to build predictive measures of motion into their informatics apparatus.

Prior to such complicated future technologies were, of course, earlier rudimentary versions of the technical racialization of appraisal. By 1925 the general program had been definitively established. One bright line indicator was a change of editorship at NAREB's NREJ. A longtime contributor, Herbert Nelson became editor with the May 18, 1925, issue. The previous editor, G. E. Henry, then began to be listed as a "managing editor" on the masthead until the August 10, 1925, issue when he relinquished all official editorial purview. Under Nelson's editorship a number of contributing editors began to populate the masthead, including Richard Ely and Ernest Fisher. With this changing of the guard came dramatic shifts in the journal.

Consider this striking contrast. In the entire year of 1924, under Henry's editorship, the *NREJ* published six articles on appraisal, with none of them explicit about race as a factor in appraisal.[102] In the first half of 1925, still under Henry's editorship, the journal published one article on appraisal, which also does not mention race.[103] In the second half of the year, now under Nelson's editorship, there appeared a full nine articles on appraisal, many of them explicit about race as a factor. In Nelson's very first issue were published two articles on the subject. One was a whimsical piece by Preston M. Nolan titled "Nolanisms for the Appraisers" compiling a set of appraisal aphorisms.[104] The second was an article by William L. Bailey of the Northwestern University Sociology Department. Bailey's "Appraising Your City" presented a "scientific" approach to the problem of the comparative evaluations of cities, "a plan known as the *rating* of communities."[105] Bailey's proposed rating involved accounting for six "fundamental economic and social factors" in community life recently made measurable by new "data."[106] Among these was the factor of "Americanization," presented as measurable by the "proportion of foreign-born over ten years of age unable to speak English."[107] The next issue featured an article on appraisal by Fisher and another by Nolan, the first in a new series of three. A few issues later, in July, there appeared an appraisal article by William Barton that does not mention race. In that same issue, however, there was published the third in Nolan's trio on appraisal, this one titled "Building Depreciation and Obsolescence."[108] Silent about race in his first two articles, as he was in his homely Nolanisms, he here turned to occasions when "an exclusive residence" is "destroyed" by what he calls "social undesirables."[109] He quickly clarified this vague term: "Values shrink in sickening fashion due to the colored invasion."[110] Later that year, Nelson's *NREJ* published another two-part article on appraisal, in which the second, "Causes of Obsolescence," included a discussion of the "shifting of urban districts" (recall Babcock's attention to city dynamics in the form of racial migration). The author, W. C. Clark, offered as an example the "nuisance" of the "Black Belt."[111]

These invective references in the latter months of 1925 were indicative of where the science of appraisal was headed. Race increasingly became a valuation factor in residential real estate as a direct cause of obsolescence.[112] That said, what these articles could not yet indicate was the extent to which obsolescence would become a category presented in the neutral and austere formats of financial valuation. Obsolescence, as the concept would develop, would have less and less traffic with explicitly moralized terminology of "undesirables" or "nuisances," and for that very reason would be able to function much more effectively to code the informatics of race into the routines of the appraisal program.

Such transformations were neither total in influence nor instantaneous in effect. Race was a contested social category in that milieu, just as it always has been in every milieu in which it has featured. Many appraisers writing books and articles did not mention race as a factor in residential real estate value.[113] Even NAREB as late as October 1927 published a typescript volume titled *Real Estate Appraising* presenting itself as a "brief course" and containing in its pages no references to race as a valuation factor.[114] It may have never occurred to some appraisers to factor race in to valuation. This is unsurprising given that up until October 1923 almost nobody counted race as a valuation factor in published work on valuation procedure and practice. And some appraisers, of course, would have explicitly resisted reading race into valuation—among this group, many would have had plain empirical concerns, while others would have had strenuous moral objections. Even at the time of its later institutionalization at government agencies in the 1930s, there was never a moment at which racialized valuation totally suffused real estate practice. But from 1923 on, it became more and more invasive.

CONSOLIDATING RACIALIZED APPRAISAL, 1931–1937

In tracing emergent assemblies of practices, conducts, and forms of subjectivity, it is misleading to search for totality. Universality in human practice is in every instance a promise. But it is not therefore a sham or a spoof. What demands our interest are the ways in which a practice prepares itself to travel anywhere and everywhere, the ways in which it can make itself seem so enticingly universalizable. The stakes for this are all the more heightened, I have been arguing, in contexts anchored by data or information—for these anchoring concepts are today packaged in a deep, full, and all-too-giddily-received promise of universality.

In the case of racialized appraisal practices, there never would be a totality. But a certain promise of universalizability was indeed tendered and widely accepted. That promise gave racialized appraisal a solidity that has still not been undone today, even after significant declines in attitudinal racism and the dismantling of some of the most obvious patterns of structural racism in the real estate industry.[115] Although racial real estate informatics were never universalized, more important is that they achieved a degree of force whereby they could be put forward as if they were uncontestable. I thus turn in this section to a series of consolidations in the first half of the 1930s.

I begin with one of the first publications emblematic of a consolidating appraisal industry: a 1931 volume published by NAREB's Appraisal Division (soon to be transformed into the American Institute of Real Estate Appraisers the following year) under the title *Real Estate Appraisals: Discussions and Examples of Current Technique.*[116] This book brought together, under the editorship of Henry A. Babcock, leading figures in the professionalization movement to provide for their colleagues a single-volume source on current appraisal techniques. It included a chapter on basic concepts of real estate value (by Philip Kniskern, soon to become AIREA's first president), a chapter on common errors of evaluation (by Frederick M. Babcock), a roster of NAREB appraisal division members organized geographically by state, a bibliography of books and articles on the subject, and a statement of professional standards for appraisers.

The volume also featured numerous elements indicative of innovative approaches to professionalization: a pair of "demonstration appraisals" in the form of actual appraisal reports with prefatory commentary by an appraiser; a series of printed blank "standard work sheets" for use in various types (residential, industrial, farms) of appraisal; and a rhetorical style by many of the volume's authors that would come to be recognized as taking the form of the genre of the "manual" guiding appraisers in their work. Throughout the volume, race is visible as an element appraisers should attend to as a factor of valuation that must be rendered into data that can be algorithmically weighted. A. P. Allingham's "Demonstration Appraisal of a Residence," reprinted the following from an actual appraisal report of a Baltimore property: "This district was opened ten years ago and is today 75% built up with a desirable class of homes of the same general character erected under rigid building and zoning restrictions. The occupancy is exclusively white and there is no undesirable element."[117] The included standardized "Residential Appraisal Work Sheet" (noted as form A-4 and available for order at $5 for one hundred copies as per an advertisement on the last page of the book) features race as a valuation category to be factored in terms of "predominant race or nationality."[118]

The greater weight given to race in appraisal practices in the 1930s can be seen by comparing Babcock's 1924 *The Appraisal of Real Estate* and his 1932 *The Valuation of Real Estate.*[119] Part 2 of the later book is titled "The Valuation Data." Here Babcock was clearly thinking through problems of nascent data management that had occupied him in the earlier book. He was now asserting that "the compilation of complete, reliable, and pertinent data is the most laborious function of the appraiser."[120] The later book was more sophisticated in its presentation of a detailed classificatory scheme of types of valuation data.[121] Babcock devoted an entire chapter to social and racial matters (titled "Influence of Social and Racial Factors on Value"). In 1924 Babcock was explicit about race as a valuation factor. But it did not appear in his chapter titles or in his descriptive table of contents. In 1932 his reader would be acquainted with race as a factor before they had even read the body of the text. In 1924 he alternated between descriptions of race as a "real" and "ultimate factor" in valuation and as one of many "other factors" in city development.[122] In 1932 race was presented as a central part of the second of "two fundamental factors."[123] His second fundamental factor was broadly described in terms of traits, characteristics, customs, and prejudices, and Babcock held that among these "racial heritage and tendencies seem to be of paramount importance."[124] The reader then confronts, in a subsection on rates of decline in residential value, the following: "Most of the variations and differences between people are slight and value declines are, as a result, gradual. But there is one difference in people, namely race, which can result in a very rapid decline."[125] Babcock's ultimate advice was a remedy "in common usage in the South"— namely, "segregation."[126]

A third indicator of the increasing force of race in valuation was a series of appraisal reports published in 1931 in the *NREJ*, still under Nelson's editorship. The Actual Appraisal Reports series was projected as a twenty-four-article showcase of a range of actual appraisals of different property types. The reports were published as an educative tool meant to model for appraisers what their work should produce.

The model reports are exemplary instances of informational outputs according to my usage of the input-process-output model of information technology. They showed how to present the information solicited, collected, stored, processed, analyzed, and computed by a congeries of information technologies. Those technologies would soon come to culminate in other forms too, such as the redlining maps to which I turn below. But preceding those more infamous neighborhood rating maps—and so historically underlying them and conceptually undergirding them—were other kinds of informational outputs.

The properties appraised in the series of model reports ranged from a California citrus ranch to a pier in New Jersey. The first piece in the series, published in January, features an appraisal of a "small store apartment and office property in outlying district of shifting values," authored by Mark Levy of Chicago.[127] The format of Levy's article is typical of the series. The piece opens with a brief narrative composed by *NREJ* editors, followed by reduced-size photographic reproductions of the sixteen actual pages of Levy's full report, including all the data types featured in those reports, from maps to filled appraisal blanks to property and building measurements to statistical tables of values to the cover letter that packaged it all together for the clients requesting the appraisal. Following a cover letter on page 1, Levy turned at

the head of page 2 to a report on "Location," in which he summarized a set
of neighborhood metrics, including number of families, grocery stores, and
banks. As the eye turns to page 3, it meets the following header in all-caps
Courier: "Predominating Nationalities." This page calls attention to the location's proximity to 26th Street, reporting that south of this street the population is 69.50 percent Negro. Levy did not comment on this. This fact about
"nationalities" is presented not in prose but as tabular data. It is as if it does
not demand commentary. The client reading the report, and by extension the
professionals eyeing it in the pages of *NREJ*, would already understand the
force of this datum. By 1931 the appearance of race in NAREB's professional
journal had consolidated. Here is race reified into data: simple givens: facts
simply set forth.

Though planned for a total of twenty-four articles, there would be twenty-
six contributions to the series before it ended in November 1932, when it
closed the loop by returning to an appraisal report of a manufacturing plant
in Chicago. Race is not mentioned in any of the appraisals of unmixed commercial property, industrial property, or other forms of nonresidential property; nor is it mentioned in every single residential appraisal. But it does appear in roughly half of the residential appraisals.

The information outputs just canvassed all format their subjects according to a series of technical standards. The subject of formatting in these appraisal reports is most obviously property. But there was an additional effort
in formatting taking place alongside of the ostensible focus. This was the formatting of people both at the level of neighborhood populations and at the
level of individual home buyers and sellers whose properties or prospective
properties were being appraised. Property and people, homes and humans
alike were rendered into racializing data by these reports. After the installation of such informatics apparatus in the 1920s, it became increasingly difficult to detach one's very self from racial data.

The sustenance that racialization drew from professionalization came not
only from new displays and visualization of data, but also just as crucially the
innovation of new technical instrumentalities. Consider that in our contemporary moment in which filing cabinets are left abandoned in favor of cloud
storage systems and in which offices sit empty in favor of timeshare mobile
work pods, it is easy to lose sight of the historical moments at which all this
now-defunct office equipment was being assembled with a veneer of ultra-
modernity. We also forget that much of that assembly involved a training in
use. A file folder does not just tell you how it wants to be used. Yet neither
does it come with instructions (at least, not anymore). A printed blank does
not ask you to fill it out. Yet we have become habituated to the technics of

forms and blanks—just as once again we are rehabituating to the usage of on-line forms that many users still find well too complicated. In the case of 1930s real estate appraisal, the movement toward professionalization was assisted by a range of equipment for burgeoning professionals who in turn came to distinguish themselves as skilled users of that very equipment.

Consider a brief three-article series by Charles Gray in the *NREJ* in 1931. According to the editor's description on the first page of the first article, titled "New Scientific System for Listing Real Estate," Gray's pieces described "a scientific record system for the real estate office."[128] Gray's article was not just a description—it was a program. It proposed a procedure to be followed that will help the professional "to clarify and to simplify the handling of his stock in trade—'his listings' and 'his prospects.'"[129] Gray's opening line advised that, "until the firm has provided proper methods, plans and equipment, it hasn't even a chance."[130] The piece recounted a bevy of procedural distinctions as the core of the business of real estate. Gray then suggested that to each procedure there corresponds an equipment: small index cards or larger file cards, all standardized in printed blanks and capable of being databased in new modern filing systems. Gray concluded his article in terms that speak directly to the object of all this professionalizing equipment: information. "It is useless to bring valuable information concerning a property into your office unless you provide adequate equipment to take care of this stock in trade. To have well kept records, full of selling information, instantly accessible is of vital importance."[131]

The penultimate section of Gray's article focused on the "manual." Manuals today often seem to us an accessory protrusion of the devices, appliances, or systems for which they provide instruction. We might be inclined to throw out the manual today, but that is only because we have already read so many manuals that we know in advance what the next one will say. They seem superfluous only to those who have already fully internalized their mechanics. In 1931, this would not have been the case. The manual was still becoming a habit. Gray underscored its importance. He noted, "The writer has thus prepared a demonstration in colors and a printed manual telling in detail just what is to be put under each heading in the forms. . . . The manual insures that the same uniformity, will prevail."[132] Record-keeping systems today are never advertised, or even noted, as coming with a manual. It would be unnecessary. We all understand that any piece of emergent equipment (such as a new software package) will come with manuals, instructions, guidebooks, and FAQs. Thus we regard the manual as an addendum to equipment rather than as equipment itself. But manuals are equipment. They instruct. They program. They specify the procedures to follow. They tell us how to achieve

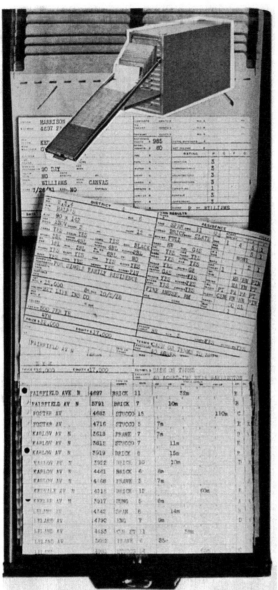

FIG. 3.1 Cards and filing apparatus from Charles Gray's "New Scientific System" for real estate agents (Gray 1931, 12)

whatever we are hoping to achieve. The manual is indeed a piece of modernizing equipment par excellence.[133]

This, at least, seems to be the case for the appraisal wing of the real estate profession. While appraisal professionalism was initially forwarded by
NAREB through its educational campaigns with their attendant technical
formats of the textbook (for new students) and the journal article (for those
already trained), the late 1920s and early 1930s saw the emergence of the appraisal manual as a technique of instruction. A number of states had been
publishing manuals for city and county assessors since the 1910s. Later on—
largely in the 1920s—these early assessors' manuals helped spawn new hybrid
forms that blended together assessment and appraisal; for instance, in the
Appraisers and Assessors Manual written by a trio of Denver assessors in 1930,
or Cuthbert Reeves's *The Appraisal of Urban Land and Buildings: A Working Manual for City Assessors,* published in 1928.[134] The emerging genre was
one of dense little books replete with valuation charts, mathematical formulae, sample forms, and foldout pages exhibiting full-size graphs of valuation
curves.

What tended to distinguish the later appraisal manuals proper from their
earlier hybrid cousins was audience. The earlier guidance documents were
for public officials. The later manuals were written expressly for emerging
real estate professionals. This latter kind of manual began to appear in the
early 1930s. A paradigmatic instance was *Boeckh's Manual of Appraisals,* written by E. H. Boeckh, of E. H. Boeckh and Associates Inc., an appraisal engineering firm. First published in 1934 by an Indianapolis company, Boeckh's
book was exceedingly technical. This is true even though much of Boeckh's
calculative apparatus was published separately (by the same publisher) in the
*Boeckh Index Calculator: For Computing Boeckh Building Cost Local Index
Numbers* (from 1936) and its replacement volume, the *Boeckh Index Calculator Tables* (from 1938).[135] These volumes—spiral bound into a plastic cover
(red spirals with black cover for my copy)—contain multipage charts (in the
1936 version) or tabular data (in the 1938 version) to assist in efficient computation of building materials (brick, lumber, cement, etc.) and construction
labor costs (carpenters, masons, iron workers, etc.). While the *Calculator*
volumes are devoted strictly to cost inputs, the *Manual* ranges much more
widely to describe methods for factoring in geography, lot size, and social factors. Boeckh's *Manual* exemplifies the way that calculations of "obsolescence"
were increasingly functioning as proxies for racialization: "A tenement may
become so depreciated that it is totally unattractive, unsanitary, and structurally unstable. . . . This is due to several things: . . . 4. The tenants may be
segregated locally because of race, nationality, or other social causes and do

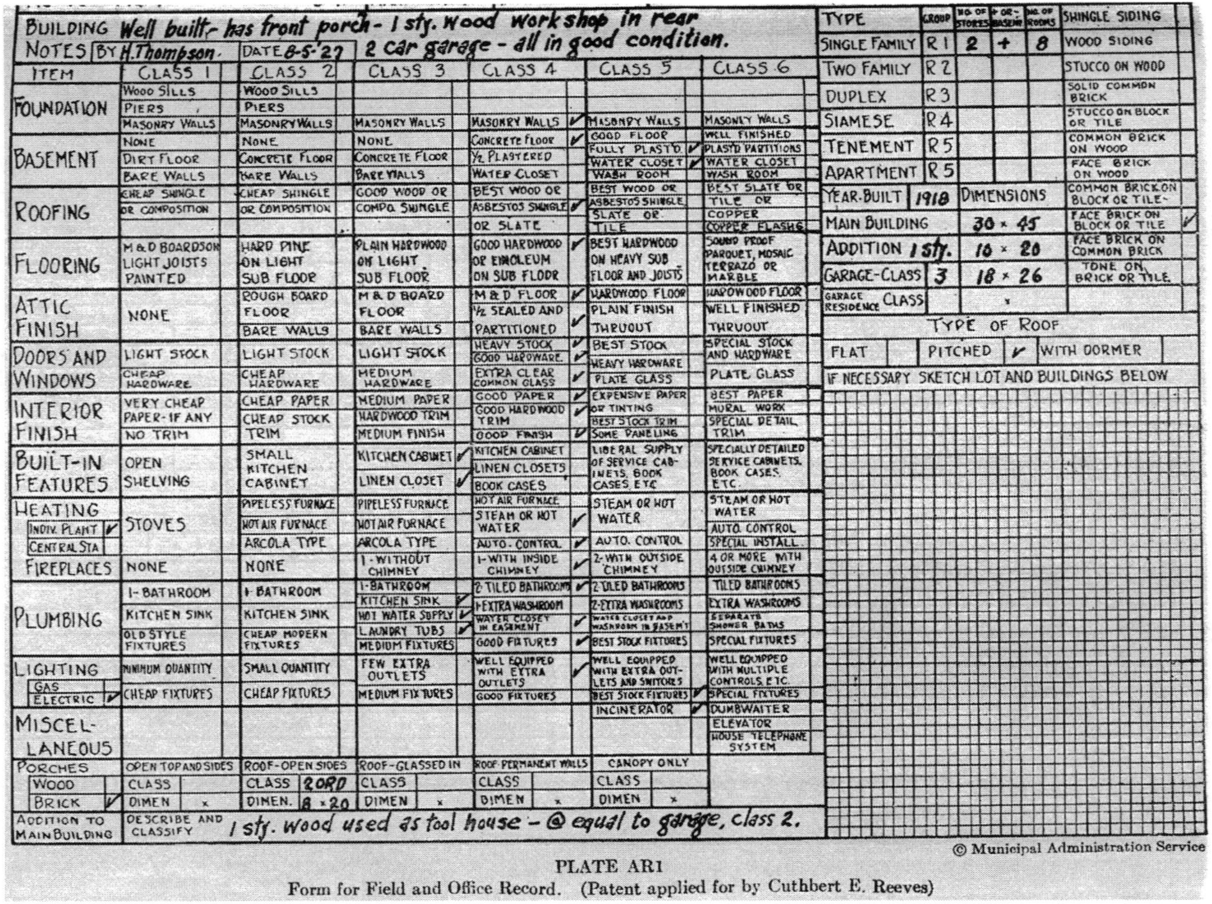

FIG. 3.2 Sample appraisal template for residential real estate (Reeves 1928, 13, plate AR1)

BUILDING Well built - has front porch - 1 sty. wood workshop in rear 2 car garage - all in good condition.
NOTES BY H. Thompson. DATE 8-5-27

ITEM	CLASS 1	CLASS 2	CLASS 3	CLASS 4	CLASS 5	CLASS 6
FOUNDATION	WOOD SILLS / PIERS	WOOD SILLS / PIERS	MASONRY WALLS	MASONRY WALLS	MASONRY WALLS	MASONRY WALLS
BASEMENT	NONE / DIRT FLOOR / BARE WALLS	NONE / CONCRETE FLOOR / BARE WALLS	CONCRETE FLOOR / BARE WALLS	CONCRETE FLOOR / ½ PLASTERED	GOOD FLOOR / FULLY PLASTRD / WATER CLOSET / WASH ROOM	WELL FINISHED / PLASTR PARTITIONS / WATER CLOSET / WASH ROOM
ROOFING	CHEAP SHINGLE OR COMPOSITION	CHEAP SHINGLE OR COMPOSITION	GOOD WOOD OR COMPO. SHINGLE	BEST WOOD OR ASBESTOS SHINGLE	BEST WOOD OR ASBESTOS SHINGLE / SLATE OR TILE	BEST SLATE OR TILE OR COPPER / COPPER FLASHING
FLOORING	M. & D. BOARDS ON LIGHT JOISTS PAINTED	HARD PINE ON LIGHT SUB FLOOR	PLAIN HARDWOOD ON SUB FLOOR	GOOD HARDWOOD OR LINOLEUM ON SUB FLOOR	BEST HARDWOOD ON HEAVY SUB FLOOR AND JOISTS	PARQUET, MOSAIC TERAZZO OR MARBLE
ATTIC FINISH	NONE	ROUGH BOARD FLOOR / BARE WALLS	M. & D. BOARD FLOOR / BARE WALLS	½ SEALED AND PARTITIONED	PLAIN FINISH / HARDWOOD FLOOR THRUOUT	WELL FINISHED / HARDWOOD FLOOR THRUOUT
DOORS AND WINDOWS	LIGHT STOCK	LIGHT STOCK	LIGHT STOCK	HEAVY STOCK COMMON GLASS	BEST STOCK	SPECIAL STOCK AND HARDWARE
(hardware)	CHEAP HARDWARE	CHEAP HARDWARE	MEDIUM HARDWARE	EXTRA GLASS OR COMMON GLASS	HEAVY HARDWARE / PLATE GLASS	PLATE GLASS
INTERIOR FINISH	VERY CHEAP PAPER - IF ANY / NO TRIM	CHEAP PAPER / CHEAP STOCK TRIM	MEDIUM PAPER / MEDIUM TRIM / MEDIUM FINISH	GOOD PAPER / GOOD HARDWOOD TRIM / GOOD FINISH	EXPENSIVE PAPER OR TINTING / BEST STOCK TRIM / SOME PANELING	BEST PAPER / MURAL WORK / SPECIAL DETAIL TRIM
BUILT-IN FEATURES	OPEN SHELVING	SMALL KITCHEN CABINET	KITCHEN CABINET / LINEN CLOSET	KITCHEN CABINET / LINEN CLOSETS / BOOK CASES	LIBERAL SUPPLY OF SERVICE CAB- INETS, BOOK CASES ETC	SPECIALLY DETAILED SERVICE CABINETS, BOOK CASES, ETC.
HEATING — INDIV PLANT	STOVES	PIPELESS FURNACE	PIPELESS FURNACE / HOT AIR FURNACE	HOT AIR FURNACE / STEAM OR HOT WATER	STEAM OR HOT WATER	STEAM OR HOT WATER
HEATING — CENTRAL STA.		HOT AIR FURNACE / ARCOLA TYPE	HOT AIR FURNACE / ARCOLA TYPE	AUTO. CONTROL	AUTO. CONTROL	AUTO CONTROL / SPECIAL INSTALL
FIREPLACES	NONE	NONE	1 - WITHOUT CHIMNEY	1 - WITH OUTSIDE CHIMNEY	2 - WITH OUTSIDE CHIMNEY	4 OR MORE WITH OUTSIDE CHIMNEY
PLUMBING	1 - BATHROOM / KITCHEN SINK / OLD STYLE FIXTURES	1 - BATHROOM / KITCHEN SINK / HOT WATER SUPPLY / LAUNDRY TUBS / MEDIUM FIXTURES	1 - BATHROOM / KITCHEN SINK / CHEAP MODERN FIXTURES / GOOD FIXTURES	2 TILED BATHROOM / EXTRA WASHROOM / WATER CLOSET WASHROOM IN BASEMENT / GOOD FIXTURES	2 TILED BATHROOMS / 2 EXTRA WASHROOMS / WATER CLOSET WASHROOM IN BASEMENT / BEST STOCK FIXTURES	2 TILED BATHROOMS / EXTRA WASHROOMS / SEPARATE SHOWER BATHS / SPECIAL FIXTURES
LIGHTING	MINIMUM QUANTITY	SMALL QUANTITY	FEW EXTRA OUTLETS	WELL EQUIPPED WITH EXTRA OUTLETS	WELL EQUIPPED WITH EXTRA OUT- LETS AND SWITCHES	WELL EQUIPPED WITH MULTIPLE CONTROLS, ETC
GAS / ELECTRIC	CHEAP FIXTURES	CHEAP FIXTURES	MEDIUM FIXTURES	GOOD FIXTURES	BEST STOCK FIXTURES	SPECIAL FIXTURES
MISCEL- LANEOUS					INCINERATOR	ELEVATOR / HOT-WATER / DUMBWATER SYSTEM

PORCHES:

	CLASS	DIMEN
OPEN TOP AND SIDES		
ROOF - OPEN SIDES	2 ORD	18 x 20
ROOF - GLASSED IN		
ROOF - PERMANENT WALLS		
CANOPY ONLY		

ADDITION TO MAIN BUILDING — DESCRIBE AND CLASSIFY: 1 sty. wood used as tool house - @ equal to garage, class 2.

TYPE:

	GROUP	NO. OF STORIES	NO. OF ROOMS
SINGLE FAMILY	R 1	2 +	8
TWO FAMILY	R 2		
DUPLEX	R 3		
SIAMESE	R 4		
TENEMENT	R 5		
APARTMENT	R 5		

YEAR BUILT 1918 — DIMENSIONS
MAIN BUILDING 30 x 45
ADDITION 1 STY. 10 x 20
GARAGE - CLASS 3 18 x 26
GARAGE CLASS / RESIDENCE CLASS

TYPE OF ROOF: FLAT | PITCHED ✓ | WITH DORMER

IF NECESSARY SKETCH LOT AND BUILDINGS BELOW

SHINGLE SIDING / WOOD SIDING — STUCCO ON WOOD — SOLID COMMON BRICK — STUCCO ON BLOCK OR TILE — COMMON BRICK ON WOOD — FACE BRICK ON WOOD — COMMON BRICK ON BLOCK OR TILE — FACE BRICK ON BLOCK OR TILE — FACE BRICK ON COMMON BRICK — TONE ON, BRICK OR TILE

PLATE AR1   Form for Field and Office Record. (Patent applied for by Cuthbert E. Reeves)

© Municipal Administration Service

not have the power of choice of habitation."[136] Boeckh, of course, is careful not to endorse this situation. Rather, he just observes it.

One of the most influential new manuals was *McMichael's Appraising Manual*, published in 1931 by Prentice-Hall; a second edition was soon issued in 1937. Both are handsome leather-bound volumes slightly smaller in size than the stiff hardback formats of the 1920s textbooks. In one's hands they feel strangely like the Bibles placed in bedside tables at motels and hotels.

Manuals like this sought to distinguish themselves in more than just material ways. McMichael described his volume as the result of an effort to "assemble and correlate many recently established facts about appraising real estate" and as a work "in both the practical and the theoretical fields of property valuation."[137] Not just a theoretical volume, the manual addressed such topics as "Physical Equipment for the Appraiser," to which an entire chapter was devoted. McMichael advised that "the appraiser can deal in accurate facts that a tapeline and careful and painstaking investigation will reveal."[138] He reproduced a photograph of a camera, a series of forms, a compass, a tape measure, and a host of other equipment—its caption read "The Real Estate Appraiser's Kit."[139] In line with Gray's recommended equipment, McMichael noted the importance for the appraiser of "card systems for indexing appraisals."[140] He also recommended filing newspaper and periodical clippings as part of the stock of data the appraiser should keep on hand. He even described how to keep these clippings: "One method for making these useful is to use heavy paper folders or pouches on which cards, properly inscribed, may be firmly attached. Make headings general in character, in order that too many files will not be required at the outset. File everything of possible interest and value, and from time to time you will be amazed to find how very useful this readily accessible fund of information is."[141] This is not quite instruction on how to use a file folder, but it almost is. It is certainly instruction in then-current data management technologies.

Beyond its recommendations of equipment and its many chapters on appraisal methods (the latter constituting the bulk of the book), McMichael's manual also addressed broader professional questions, including codes of ethics. NAREB had devoted attention to its industry-wide code in 1924. But it waited until 1929 to adopt a code covering appraisal practices. Continuing his earlier efforts on behalf of professionalization from 1923, McMichael's 1931 manual reprinted this code in full (to a length of fourteen pages) and recounted brief portions of its history.[142] The volume clearly aimed to serve as a one-stop reference for the appraiser conducting his work.

Just as McMichael's manual in multiple ways furthered his previous work toward professionalization, he also used his manual to extend his previous commitments to the racialization of appraisal. Chapter 26 is titled "Appraising Property in "Twilight Zones or Blighted Areas" and takes up the difficult work of valuation in what were then coming to be called "slums."[143] The chapter is forthright about the role of race in obsolescing areas. McMichael quoted one author discussing such areas as caused in part by "infiltration of a lower social type."[144] He then quoted another, who describes residential blight as partly caused by "invasion by incompatible new residents."[145] Both of these phrases are embedded in long quotations—neither mentions race explicitly. McMichael invoked them on the way to a presentation, in the form of a table, of the primary and secondary causes of blighted areas. One item listed under "Primary Causes" is "3. Invasion by incompatible uses."[146] This item is further subdivided into five classes, including first, "a. Social or racial changes," and penultimate and last, "d. Heavy traffic," and "e. Smoke or smells from near-by areas."[147] Six pages later this trio is echoed. The context here is chapter 28, "Some Factors to Observe in Making an Appraisal," in which is offered a four-page list of questions whose answers will supply "the information that a valuator should obtain."[148] The list does not admit of the neat subdivisions of McMichael's slum appraisal table—there are only three broad headings for a list of well over sixty questions. Under the "In General" heading, McMichael advises appraisers to answer some thirty questions. Two of them, printed one right after the other, are as follows:

> Are there undesirable racial elements in the neighborhood, and, if so, are they likely to expand in a way that may injure the property? Are there other undesirable factors, such as odors from stockyards, smoke from factories, odors from swamps, overflows from watercourses, undue noise from passing trucks, exhaust fumes from great numbers of automobiles, etc., etc., which may reflect on future use and value?[149]

The combination is jarring. However, difficult to notice if we attend only to the attitudinal racism we would expect ("undesirable" as attributed to race alongside swamp odors and traffic exhaust) is the simple injunction to factor "racial elements" into the appraisal. This chapter, one of the last in the book, offered its reader a kind of code for the measure of worth. The appraiser who answered each question on the list would have followed the procedure to the scientific answer. This, precisely, is how race was factored into valuation as one of many elements featuring in an algorithm.

The 1937 edition of *McMichael's Appraising Manual* introduced additional material on these topics.[150] It retained almost verbatim the table of primary and secondary causes of blight as well as the list of factors to observe. But it appended to the chapter on residential appraisal a pair of "interesting tables"

ADDITIONS	Per Cent
Site.	15
Type of neighborhood and social factors.	20
View and climate.	15
Public utilities and schools.	5
Streets and alleys; distance to work in city.	10
Contour and soil.	5
Physical environment.	20
Restrictions and planning.	10

A table of common deductions reducing the above percentages of value follows:

DEDUCTIONS	Per Cent
Noise and dirt, up to.	25
Racial and foreign neighbors, up to.	60
Adjacent vacancy, up to.	20
Poor architecture, up to.	20
Obsolescence, up to.	70
Distances from city, work, schools, etc., up to.	100
Nuisances (funerals, freight, trucks, etc.), up to.	100
Dead-end streets, up to.	15

FIG. 3.3 Table of common deductions (McMichael 1937, 330)

given to McMichael by a Thurston Ross of Los Angeles. Both are tables of value weights: one a table of value additions and the other a table of value deductions. The latter "table of common deductions" proposes a "racial and foreign neighbors deduction up to 60%."[151]

If McMichael in 1931 had sought to treat race as part of a list alongside other such factors as the presence of nearby public services or of nearby vacant lots, here in 1937 he presents the possibility of race as a specifiable numerical factor in a valuation algorithm. This is a clear example of the factoring of race within a stepwise procedure, or what Babcock had presciently called a "program" in his 1924 book. These algorithmic treatments of racial data were often accompanied by seemingly racist descriptors. But even without the accompaniment of the benighted morals that fund McMichael's rhetorical associations, we can take notice of the very presence of race as an item in a table of deductions. That presence persisted even in the absence of any benighted morality funding it. And that is much more chilling.

It soon became increasingly rare to find overtly racist judgments in appraisal manuals. In part this was a function of the congealing of the form of the manual. One would not expect to confront overt racism in a manual today. What book of instructions endorses, judges, or moralizes? Imagine how

odd it would be to find a line in your refrigerator manual endorsing or con-
demning a certain culinary preference, let alone some cultural food practice.
Such rhetoric seems radically inappropriate for the genre. Anticipating what
we expect of our manuals today, the appraisal manuals merely restate as fact
that race is a factor in residential real estate valuation.

What is dangerous in this is the comforting assumption that midcentury
racism came to a close when midcentury attitudes were swept away by late-
century tolerance. This shift to liberal tolerance was, I would argue, a good
thing—indeed, a very good thing. But as an attitudinal shift it could hardly
have been sufficient to address embedded technologies of racism. Insofar as
the informatics of race remains installed in all kinds of projects and is deployed much more widely than just in residential real estate valuations, we
continue to be subject to a technological racism effect. The lingering history
of government redlining shows why.

LEVERAGING THE INFORMATICS
OF RACE IN THE STATE, 1933–1938

After being consolidated by real estate professionals and their professional
associations, the racializing informatics of appraisal migrated into the key
institutions of federal housing policy. These included the Federal Home Loan
Bank Board, the Home Owners' Loan Corporation (HOLC), and the Federal
Housing Administration (FHA). At the HOLC and the FHA especially, the
consolidating work of NAREB and AIREA would become deeply entrenched.
This was not just the work of a sovereign state power lending its force to
action at first organized elsewhere. It was also a moment when a power of
information invaded, made use of, and transformed the very mechanisms
of the state. From an institutional perspective, the HOLC and the FHA may
look like so many other state agencies. But from a technical perspective, they
were a hotbed of transformation. What this suggests is that even where more
overt aspects of institutional racism at such federal agencies were removed,
there could remain an ongoing technological racism of the algorithm that
persists unaddressed because unacknowledged. Once federal agencies began
to explicitly dismantle their racisms in the 1960s, they may have yet left intact
a functioning informatics of race that could be used to operate a technological racism. This form of racism still lacks its critics, its opposition, and its
resistance—in part because it lacks its critical diagnosis.

The HOLC began producing redlining maps in 1933. These city maps were
visual representations of data analytics performed on urban racial data. In
many instances, the data underlying the maps was collected by the HOLC,

though there were also plenty of cities for which the data had already been created by the offices of city assessors, county registers, or city and county engineers. In addition to the data were the algorithms. The analytics used by the HOLC in many of their aspects simply reproduced the data processing work that had been developed by professional appraisers in the decade or so prior. As such, the specifically technological racism of the state was not at bottom a state product.

To see why this matters, we need to zoom out to the broader context within which these maps were produced. Homeownership was one of the most important economic victims of the Depression. President Herbert Hoover recognized this early on when in 1931 he convened the President's National Conference on Home Building and Home Ownership. But this group's recommendations proved ineffective. For example, the 1932 Federal Home Loan Bank Act established a reserve of credit for mortgage lending in an attempt to invigorate the housing market. Approximately forty-one thousand Americans applied for these loans in the first two years of the program. A total of three applications were approved.[152] When Roosevelt came into office in March 1933, he pursued a more proactive approach. The Emergency Farm Mortgage Act was passed in May. In June there followed the establishment of the HOLC. Just over a year later, the National Housing Act would establish the FHA.

The significance of the impact of the HOLC on American homeownership is hard to overstate. Prior to the HOLC, homeownership was widely inaccessible. Mortgages were a relatively new financial instrument that had been introduced into wide usage only in the late 1910s. Through the 1920s they remained for the most part short-term balloon mortgages of five or ten years, at the end of which most buyers would have to seek a new mortgage to pay off the remaining balance (or the balloon) on the previous mortgage. Minimum down payments were typically around 30 percent of the value of the home. The HOLC began to change all of this as part of a program of modernizing mortgage markets. The HOLC's main contribution was the long-term, self-amortizing mortgage. Repayment periods were extended to twenty years, at the end of which borrowers would be completely paid off rather than needing to seek a second mortgage for a balloon payment.

The HOLC began accepting applications in June 1933 and stopped accepting them in June 1935. They received applications on approximately 40 percent of all eligible properties nationwide (these were nonfarm, owner-occupied, one-to-four-family dwellings).[153] Of these applications, almost half were rejected or withdrawn.[154] But for the 1,017,821 applicants who were accepted, the HOLC essentially took over their mortgages in order to give the borrowers better terms.[155] They did so by issuing low-interest bonds to lenders (typically banks, savings and loans, and insurance companies) in exchange for the mortgage. Lenders were inclined to take the deal on the rationale that a guaranteed low-interest payment was more lucrative than receiving higher-interest payments from impoverished homeowners unlikely to make their payments. The HOLC then constructed, held, and serviced new longer-term and lower-interest (and therefore lower-installment) loans.

In addition to modernizing mortgage instruments in a way that increased the accessibility of credit, the HOLC also helped to standardize appraisal practices by virtue of the huge volume of appraisals it was conducting. As Jackson has noted: "The HOLC systematized appraisal methods across the nation. . . . The element of novelty did not lie in the appraisal requirement itself—that had long been standard real-estate practice. Rather, it lay in the creation of a formal and uniform system of appraisal, reduced to writing, structured in defined procedures, and implemented by individuals only after intensive training."[156] This formalization, moreover, borrowed many of its standards from the then-new industry practices canvassed above.

The bill that brought the HOLC into being did not specify an appraisal procedure to be used by the agency. But one would be needed. An early publication by the agency titled *Loan Regulations*, produced for the benefit of its employees devoted a brief section to appraisal.[157] It stated that HOLC appraisals should seek "the fair worth" of a property, which was not necessarily equivalent to "the technical market value."[158] It was proposed that fair worth could be analyzed as an equally weighted combination of market value, reproduction cost to build a similar building on a similar lot less depreciation, and value of the premises as a function of a reasonable expected rental income. According to one early historian of the HOLC writing in 1951, such a "proposal to use a formula was, in fact, a rather novel procedure in appraising residences."[159] But we have seen that this is not strictly true. Formulaic appraisal procedures were already well established by the second half of 1933. Indeed, it would have been a surprise for the HOLC not to employ a formula of some kind, for, as the same historian notes, the HOLC consulted in late 1933 with AIREA in preparing regulations on qualifications for appraisers it would hire (either as salaried employees or on a fee basis).[160]

It would not be long before the HOLC would produce its own standardized appraisal form, in January 1934.[161] The form—four pages long and covering ninety-eight items—focused mostly on physical characteristics of the dwelling. It also indicated a need for data on the racial composition of neighborhoods as either "native white," "native white and foreign," or "Negro."[162]

This standardization of the racialization of appraisal has been understandably overshadowed by another racial technology produced at the HOLC: the infamous redlining maps. As part of its efforts, the HOLC compiled a huge number of city survey files on American urban areas. One element in these files were the residential security maps, or what are now colloquially known as redlining maps. These large maps were the visual yield of an elaborate sequence of data collection and information-processing technologies. The maps presented large sections of municipalities divided into numerous sections, each of which was graded on a scale of "residential security." There were four grades; the fourth was coded by a *D* and on most maps was shaded in red. That shading explains the otherwise opaque term "redlining" that functions as a euphemism for the full informatics apparatus of real estate discrimination. The maps were rediscovered by Jackson in his work on Record Group 195 at the National Archives in preparation for his groundbreaking 1985 book *Crabgrass Frontier*. He noted there that "black neighborhoods were invariably rated as Fourth grade," or security grade *D*, or redlined.[163]

Jackson first reported on the maps in a 1980 article in *Journal of Urban History*.[164] Much has been made of these cartographic surveys in the years since. Recent research by Amy Hillier casts some doubt on the idea—more an inference made by many who have since heard about the maps than an assertion actually verified—that redlining was a widespread and influential practice.[165] Hillier worries about the lack of empirical connections that would support the inference from these mappings of racial data to actual racist lending patterns: "By themselves, maps with red lines are not adequate proof of redlining."[166] This caution is in keeping with Jackson's own conclusion that "strong evidence indicates that HOLC did in fact issue mortgage assistance impartially and make the majority of its obligations in 'definitely declining' or 'hazardous' neighborhoods."[167] Hillier's point is not that there was no credit market racism in this period. Rather, she merely calls into question the claim that redlining maps were a key cause of credit market racism. Her primary evidence is that the maps were not widely distributed.[168] Yet there was credit market racism. Thus there must have been other techniques of racialization that functioned to substantiate racism in the credit markets. Hillier concludes: "A variety of evidence suggests that HOLC was not responsible for redlining."[169]

Here is where the work of racialized appraisals could have been more important than previously recognized. The racialized informatics of real estate appraisal preceded the maps, the HOLC, and the FHA. Once these agencies were created they did not need to invent new appraisal methods. Rather, they could have, and most likely would have, adopted the appraisal practices that had already been circulating for almost a decade by NAREB appraisers, editors, and authors. One fact suggesting that they may have in actuality adopted

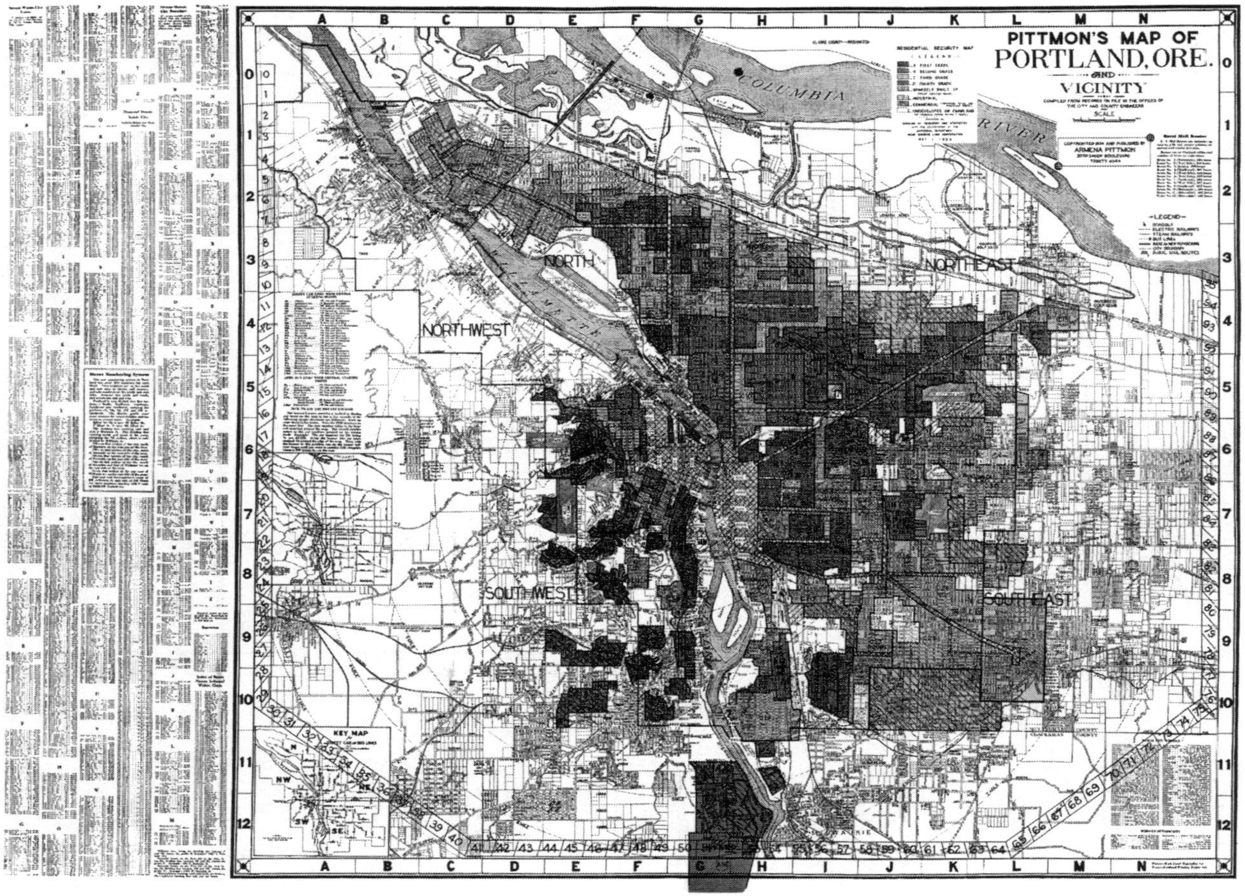

FIG. 3.4 HOLC redlining map for the city of Portland, Oregon, produced in 1934 (R. Nelson et al., n.d.)

those methods is that the same people who had crafted them for industry were later employed as appraisal officials at these agencies. For example, one of the primary agents of both the professionalization and racialization of appraisal in the 1920s later became one of the architects of the appraisal practices of the FHA as embodied in its *Underwriting Manuals*: Frederick Morrison Babcock, chief underwriter for the FHA.[170]

Both the FHA and the HOLC appeared to have factored race into appraisal. But only the former seems to have exhibited discriminatory lending patterns—a paradigmatic form of institutionalized racism. As Jackson wrote of the FHA, "For perhaps the first time, the federal government embraced the discriminatory attitudes of the marketplace."[171] In what way did it do this? At least two. First, and obviously, was endorsement of the racism of racial covenants, which lasted until 1950.[172] Second, and less obvious, was deployment of appraisal practices that explicitly factored race into market values for residential properties.

FHA's appraisal standards cannot be regarded as straightforward mandates for racial discrimination, unlike the agency's open endorsement of discriminatory covenants from its inception in 1934 until 1950 (two years after they were forbidden by *Shelley v. Kraemer*). Appraisal standards are not laws, ordinances, or contracts that compel. Rather, much more subtly, they are guidelines that, in the years under survey here, functioned as an invitation to the installation of an informatics of race on whose basis racial discrimination could and would travel.

There is good evidence that FHA guidelines built race into its appraisal algorithms. The "Report of Valuator" form appended at the end of the 1938 edition of the *Manual* is a specimen of a printed blank that would have seen everyday use by FHA appraisers. It does not itself contain any direct references to race. The form's "Rating of Location" table on the reverse side contains eight criteria of valuation, each to be given a rating of 1 to 5. These include "Relative Economic Stability" and "Protection from Adverse Influences" in language that already echoes discussions of race elsewhere in the *Manual* (the *Manual* being only the instructions for this and other forms). According to these instructions, the eight criteria are numerically weighted to compute a location-rating score referred to as "Total Rating of Location."[173] Each of the eight categories is accompanied not only by five columns in which each score's weighted value is indicated, but also by an initial column labeled "Reject" in the case that a location rates so low in a specific category that the property ought to be rejected outright.[174]

The first category of "Relative Economic Stability" is a two-part rating that involves both the overall economic background of the area, and then a rating of the specific location vis-à-vis its area. The former is more citywide and so is

2015—Report of Valuator

Rating of Location (Established Rating of Location used for comparison—No. ....)

FEATURE	REJECT	1	2	3	4	5	RATING
Relative Economic Stability							
Protection from Adverse Influences		4	8	12	16	20	
Freedom from Special Hazards		1	2	3	4	5	
Adequacy of Civic, Social, and Commercial Centers		1	2	3	4	5	
Adequacy of Transportation		2	4	6	8	10	
Sufficiency of Utilities and Conveniences		1	2	3	4	5	
Level of Taxes and Special Assessments		1	2	3	4	5	
Appeal		2	4	6	8	10	
TOTAL RATING OF LOCATION							

Rating of Property (Assuming proposed or required repairs or alterations have been made)

PHYSICAL SECURITY FEATURES	REJECT	1	2	3	4	5	RATING
Durability — Structural Soundness		5	10	15	20	25	
Durability — Resistance to Elements		2	4	6	8	10	
Durability — Resistance to Use		1	2	3	4	5	
Function — Livability and Functional Plan		4	8	12	16	20	
Function — Mechanical and Convenience Equipment		2	4	6	8	10	
Function — Natural Light and Ventilation		2	4	6	8	10	
Architectural Attractiveness		4	8	12	16	20	
Total Rating of Physical Security							
Adjustment for Nonconformity		12	9	6	3	0	
TOTAL RATING OF PROPERTY							

(17) The characteristics of the immediate neighborhood are indicated as follows:

a.　☐ Part of metropolitan area.　　☐ Close-in.　☐ Outlying.
　　☐ Isolated community.　　　　　☐ Expanding population.
　　☐ Partly built-up ....%　☐ Stationary population.
　　☐ Built-up ....%　☐ Declining population.

b. Location ☐ is; ☐ is not in an Undeveloped Subdivision.
c. Typical family annual income level. $..........................
d. Rental spread of typical properties.. $.............. to $..........
e. Price range of typical properties...... $.............. to $..........
f. Typical age of main buildings .... years.
g. Is change in class of occupancy ☐ occurring, ☐ threatening, ☐ remote?
h. Name existing or threatening adverse influences ....
....
....
....
....

i. ....% owner-occupancy.　j. ....% homes vacant.
k. Indicate present demand for dwelling properties:
　　☐ None. ☐ Little. ☐ Moderate. ☐ Strong.

(18) Compared with other properties in the immediate neighborhood will this property at the price reported as its value be—
　　☐ More salable? ☐ Equally salable? ☐ Less salable? Why?
....

(19) Remaining economic life of building .... years.
(20) Owner-occupancy appeal ....%
(21) Monthly rental value, unfurnished................... $..............
(22) Amenity increment ....% $..............
(23) Derived monthly value........................... $..............
(24) Conversion factor ....
(25) Valuation of excess land .... $..............
(26) Derived capital value .... $..............
(27) Total replacement cost of property.................... $..............
(28) Available market price .... $..............

ESTIMATE OF VALUE.—In my opinion the value of the property described above, assuming the contemplated improvements or new construction described in exhibits, if any, accompanying FHA Form No. 2004a, or assuming the repairs or alterations or additions, if any, listed under item (15) have been completed, is...

Distribution of value estimate: Land.................... $.................... @ $.................... per ☐ lot, ☐ fr. ft., ☐ sq. ft.
　　　　　　　　　　Main Building.................... $....................
　　　　　　　　　　Garage.................... $....................
　　　　　　　　　　Other Improvements.......... $....................

FIG. 3.5 Portion of the "Report of Valuator" form from the FHA's *Underwriting Manual* (US Federal Housing Administration 1938, appendix)

established by a chief valuator, and the latter is specific to the location and the responsibility of the valuator filling out the form for that location. Among the qualities taken into account at the location level are "social characteristics of neighborhood occupants," including "the moral qualities, the habits, the abilities and the social, educational and cultural backgrounds" of neighborhood residents."[175] The second category was "Protection from Adverse Influences." Both the 1936 and the 1938 *Manual* contain a subsection here (titled in the latter manual "Quality of Neighborhood Development") in which it is asserted (again, in the language of the latter, though it is nearly identical in both):

> Areas surrounding a location are investigated to determine whether incompatible racial and social groups are present, for the purposes of making a prediction regarding the probability of the location being invaded by such groups. If a

neighborhood is to retain stability, it is necessary that properties shall continue
to be occupied by the same social and racial classes. A change in social or racial
occupancy generally contributes to instability and a decline in values.[176]

The manuals present race as an item of data that begs to be crunched by a robust informatics engine. Racial homogeneity, the absence of "incompatible"
groups, and the lack of a threat of "invasion" are all factors to be computed.
Race had been formatted into valuation.

Consider now another manual produced by the FHA in a joint project with
the Works Progress Administration: the 1935 *Technique for a Real Property Survey* (*TRPS*).[177] According to the FHA's *Second Annual Report* on agency activities, the goal of the *TRPS* was "a standard procedure for conducting comprehensive real property surveys."[178] In short, the survey was intended to produce
a mass of "raw" property data that could later become a basis for appraisal
reports, redline maps, and other au courant information products. Some of
those data were collected via the *TRPS*'s "Dwelling Schedule" that enumerators
would take door to door.

Box N of the "Dwelling Schedule" was a site for the informationalization
of race. At the doorstep, after a short greeting and conversation, what was
in one moment the visibility of race—phenotypical appearance—became in
the very next moment an informatics of race—a tick in box N according to
a protocol: "N. Race of Household: 1. White, 2. Negro, 3. Other." In this case
the protocol was: "Mark in the appropriate box the race of the household,
whether white, negro, or other. If any member of the household, other than
servant, is negro or of a race other than white, consider the whole household
as belonging to that race."[179]

Whatever specific information processing the Dwelling Schedule man-
dated—for instance, whether it allowed mixed race at the level of household
or individual—there was already a politics at play in the very transformation
of race into data. There was already an exercise of power whenever box N arrived at the doorstep to do its work of formatting racial identity. Here in this
little box, on this little form, is thus consolidated an entire sociotechnological
assembly for the production of an informatics of race. Latour observes that
"this operation of ticking rows and columns with a pencil is a humble but crucial one."[180] Box N is an instructive example of the powerful work of formatting that has been my focus throughout this book. It is impossible for social
practices of racialization to become instantiated in technosocial practices of
data without tiny technologies like box N. And once such technologies are
put into operation, they silently perform their work of reformatting social acts
of racialization into scientific processes of datafication. Race is thereby made

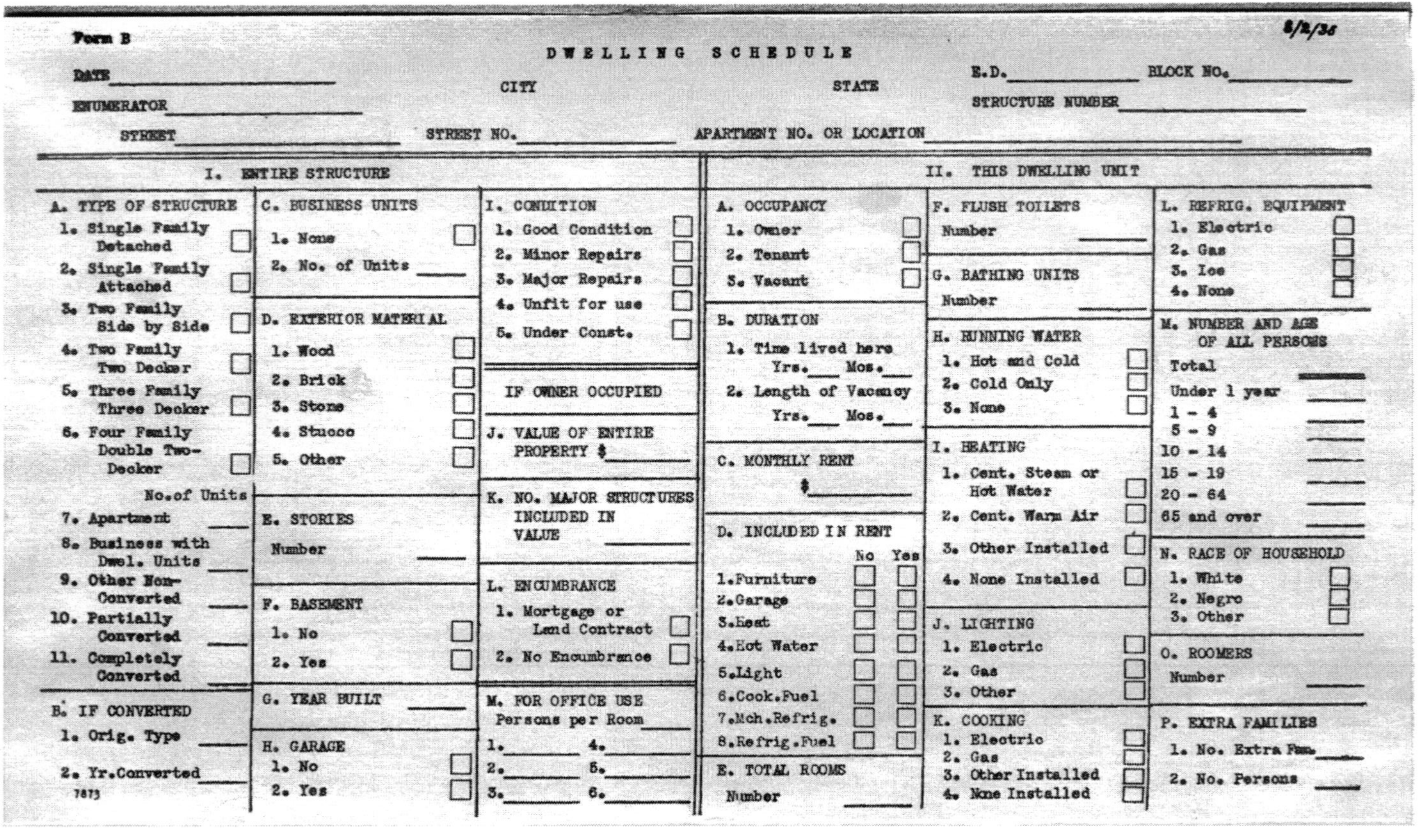

Form B　　　　　　　　　　　　　　　　　　　　　　　8/2/35

DWELLING SCHEDULE

DATE _____

ENUMERATOR _____

STREET _____　STREET NO. _____　CITY _____　STATE _____　APARTMENT NO. OR LOCATION _____　E.D. _____　BLOCK NO. _____　STRUCTURE NUMBER _____

I. ENTIRE STRUCTURE

A. TYPE OF STRUCTURE
1. Single Family Detached
2. Single Family Attached
3. Two Family Side by Side
4. Two Family Two Decker
5. Three Family Three Decker
6. Four Family Double Two-Decker

No. of Units
7. Apartment
8. Business with Dwel. Units
9. Other Non-Converted
10. Partially Converted
11. Completely Converted

B. IF CONVERTED
1. Orig. Type
2. Yr. Converted

C. BUSINESS UNITS
1. None
2. No. of Units

D. EXTERIOR MATERIAL
1. Wood
2. Brick
3. Stone
4. Stucco
5. Other

E. STORIES Number

F. BASEMENT
1. No
2. Yes

G. YEAR BUILT

H. GARAGE
1. No
2. Yes

I. CONDITION
1. Good Condition
2. Minor Repairs
3. Major Repairs
4. Unfit for use
5. Under Const.

IF OWNER OCCUPIED

J. VALUE OF ENTIRE PROPERTY $

K. NO. MAJOR STRUCTURES INCLUDED IN VALUE

L. ENCUMBRANCE
1. Mortgage or Land Contract
2. No Encumbrance

M. FOR OFFICE USE Persons per Room
1.　4.
2.　5.
3.

A. OCCUPANCY
1. Owner
2. Tenant
3. Vacant

B. DURATION
1. Time lived here Yrs. ___ Mos. ___
2. Length of Vacancy Yrs. ___ Mos. ___

C. MONTHLY RENT $

D. INCLUDED IN RENT　No　Yes
1. Furniture
2. Garage
3. Heat
4. Hot Water
5. Light
6. Cook-Fuel
7. Mch. Refrig.
8. Refrig. Fuel

E. TOTAL ROOMS Number

II. THIS DWELLING UNIT

F. FLUSH TOILETS Number

G. BATHING UNITS Number

H. RUNNING WATER
1. Hot and Cold
2. Cold Only
3. None

I. HEATING
1. Cent. Steam or Hot Water
2. Cent. Warm Air
3. Other Installed
4. None Installed

J. LIGHTING
1. Electric
2. Gas
3. Other

K. COOKING
1. Electric
2. Gas
3. Other Installed
4. None Installed

L. REFRIG. EQUIPMENT
1. Electric
2. Gas
3. Ice
4. None

M. NUMBER AND AGE OF ALL PERSONS
Total
Under 1 year
1 - 4
5 - 9
10 - 14
15 - 19
20 - 64
65 and over

N. RACE OF HOUSEHOLD
1. White
2. Negro
3. Other

O. ROOMERS Number

P. EXTRA FAMILIES
1. No. Extra Fam.
2. No. Persons

7873

FIG. 3.6 "Dwelling Schedule" from *Technique for a Real Property Survey*, form B, part 1, in the section "General Procedures" (US Federal Housing Administration/Works Progress Administration 1935, 14) data. It is not thereby unmade in its other forms. The unseen data of race are additive. The infopolitics of race is thus layered on phenotypes, biologies, bodies, and more. The result is that racialization thereby became different, more multiple, more heterogenous, and therefore more complicated to track.

It is crucial to underscore, then, that the informatics of race enacted by these technologies were not merely offered as optional for appraisal technicians. Rather, the FHA built them into the requisite mechanics of their information technology. Their valuators and enumerators were instructed to take them into account.

By thus building race into its appraisals, the FHA would have a profound effect on racializing homeownership—and a much deeper effect than the HOLC. Although the HOLC assumed about one million loans in a few short years in the mid-1930s, the FHA itself was not in the business of lending or of taking over defaulted mortgages. Rather, the FHA insured the long-term mortgage loans made by private lenders. To qualify for this insurance, loans had to meet certain conditions. The FHA's terms for loans went even further than the HOLC's in extending the life of loans (from twenty to thirty years), decreasing the size of down payments (to less than 10 percent), and lowering interest rates (by an additional two percentage points). FHA appraisal standards—which had first filtered into the agency by way of appraisal professionals like Babcock—would be pushed out into the private lending market by the very nature of the arrangement, for the promise of FHA guarantees gave lenders a huge incentive to seek loans that met their standards. FHA practices thus became nearly ubiquitous. Again, they were not universal; but they were certainly universalizable throughout the home finance market. As standards, this is exactly what they were meant to be.

"The Racial Data Revolution"

The informatics of racialized real estate appraisal is part of a much broader program through which a new kind of racialization was coded in the early decades of the twentieth century. Khalil Gibran Muhammad refers to the emergence of this program as "the racial data revolution."[181] As he describes its stakes, "Racial knowledge that had been dominated by anecdotal, hereditarian, and pseudobiological theories of race would gradually be transformed by new social scientific theories of race and society and new tools of analysis, namely racial statistics and social surveys."[182] I have here interrogated just one dimension of this datafication of race, which is itself just one dimension of a broader datafication of the self that was being engineered in these decades. The other dimensions are not unimportant. Consider just one as an index of the broader history at play here: Muhammad's narrative of the production of black criminality in roughly the same times and places.

One point of culmination in the production of black criminality was the introduction of the Uniform Crime Reports (UCR) in 1930. These reports, which are maintained to this day, were developed by a diverse constituency of actors ranging from the Federal Bureau of Investigation to the Social Science Research Council. Lawrence Rosen notes in his history of the UCR that the particular conjuncture that resulted in this "first nationwide crime data system in the United States" was actually "not the first attempt to introduce a national system of measuring crime."[183] What made the UCR different and what made it successful was the uniformity it imposed on data reporting. The issue of uniformity, claims Rosen, was "the most difficult problem in the creation of a national crime data system in the United States."[184] The 1930 UCR thus contrasts with earlier eighteenth- and nineteenth-century productions of criminal informational dossiers, such as those discussed in Foucault's *Discipline and Punish*.[185] It is easy to be impressed by all those earlier files. But we must bear in mind the provincialism that inflected all that early paperwork. Those dossiers did not become a mode of power in their own right until they were finally formatted in concert. Paper presses itself upon us only when it is standardized in ways that can transcend provinciality and assert itself as the ubiquitous universal of information. The emergence of the UCR in the 1920s and 1930s marks that moment of mass formatting for criminal records. The calls for uniformity culminating in the UCR were calls for second-order data, or metadata that formats first-order data. Through such metadata the UCR made criminality into a constant quality whose variations could be reliably measured.

The technical apparatus at play here is exemplified in the work of a technician featured in Muhammad's history of the data revolution in black criminality. Sociologist Thorsten Sellin in 1928 was a leading figure in the fight "to disentangle race from crime."[186] Sellin would not be entirely successful in this. Rather, on Muhammad's account, he was successful in differently entangling race and criminality. What matters for my account is simply that whatever success Sellin met with depended on data. He did not unlink race from crime by untying either from data. Rather, Sellin's arguments relied on data. A 1928 paper cited by Muhammad shows Sellin as someone who makes his argument primarily through the evidence of tabulated statistics.[187] Two decades later, in a 1950 paper, Sellin discussed the UCR as a project of data about data: "It would be impossible to form any valid opinion about the amount of criminality in a given jurisdiction nor could we know how criminality changes over a period of time or how its component parts vary, were it not for the fact that information in the possession of police and other agencies can be tabulated and analyzed."[188] The uniformity of data is for Sellin the key to what the UCR promises, but fails, to be. Sellin's arguments for delinking race from criminality depended in part on a reckoning with the formatting work of metadata. Such arguments take place on the terrain of what I have been calling infopolitics.

Looking forward to today from these and other formatting wars of the early twentieth century, we can raise a number of questions. Are technologies of racializing informatics still with us? Are the technological racisms that rely on them intact? And why does the informatics of race continue to matter?

First, these technologies are undeniably still with us. Think of the thousands of boxes you have ticked in your lifetime beside which were written the basic words composing the American racial taxonomy. Think of the databases into which those ticks have flowed. Think of the algorithmic production of correlations built on these databases upon databases. These tiny technologies—and the much vaster assemblies they are enrolled in—are still with us. And they continue to be partially productive of race in at least two ways. First, they are the basis for all manner of political programs and social projects that hang upon racial data. Further, they are ongoingly productive of the datafication of racial identity. Race is practiced in a multiplicity of ways. Among this multiplicity, it is, for us today, going back to the 1920s, practiced through personal data. Every time we tick the box, or ask another to tick it, we confront the datafication of race, which is to say that we confront racialization in the form of data. We do not all confront informationalized racialization in the same way. But precisely because we all much confront it in such unceremonious forms as printed blanks and digital forms, it can be turned into a technology of differential impacts.

This raises my second question: Are these techniques still used for racist purposes? To the extent that the informatics of race was put to use for racist purposes in the middle decades of the century, and to the extent that this technological racism still today remains largely unexamined and even unacknowledged, it would be shocking if such racisms have disappeared. Recent research indeed shows that the home finance industry today, despite the purported neutrality of its creditworthiness algorithms, continues to be a site of effective even if often unintentional racial discrimination.[189] In a different domain, some of our newest uses of racial data—for instance, in machine-learning projects seeking to replicate human intelligence in code— exhibit unexpected patterns of racial discrimination.[190] In both cases, these bad outcomes may be an effect of algorithms that rely on designs in which are embedded the racist legacies for which racializing informatics have so often been deployed.

Can we now say why the informatics of race matters? What we can say is that it matters because the politics of race is now not only a politics of bodies and bloodlines, but also a politics of algorithms and analytics, of data and documents. Quietly tucked away in rack-mounted servers there lie masses of data waiting for future uses both known and unknown. Without those data, there would be no uses. Without those uses, there would be different conceptions of who we are. But this does not mean that destroying the data of race could possibly end racism. All it suggests is that the specific formats of those data are very much operative in the contemporary politics of race. If James Baldwin was right to say that "people are trapped in history and history is trapped in them,"[191] then this would be no less true for informational people. We are all pent up in the history of the deployment of racializing informatics.

PART II

Powers of Formatting

Toward a Political Theory for Informational Persons

Who We Are

In his memoir *Black Boy*, Richard Wright recalls coming of age in Mississippi before he made his way up to Memphis and then finally to Chicago in 1927, one of millions of southern black Americans moving north during the Great Migration. Wright powerfully conveys how racism can exert its brutality in forms that appear humbly unremarkable.

One of the book's numerous episodes stands out as exemplifying the changing techniques through which power began to operate in the early twentieth century. What it specifically illuminates is how then-emerging information technologies facilitated the perpetuation of an old politics of race through new means. The scene is one in which Wright tells of an injustice doled out to his maternal grandfather, Richard Wilson. Born into slavery in 1847, Wilson escaped during the Civil War and joined the Grand Army of the Republic, in which he was injured during service. He died in 1921 having never received a single payment on the disability pension owed him by the federal government. Wright recounts:

> In the process of being discharged from the Union Army, he had gone to a white officer to seek help in filling out his papers. In filling out the papers, the white officer misspelled Grandpa's name, making him Richard Vinson instead of Richard Wilson. . . . Grandpa did not discover that he had been discharged in the name of Richard Vinson until years later; and when he applied to the War Department for a pension, no trace could be found of his ever having served in the Union Army under the name of Richard Wilson. . . . For decades a long correspondence took place between Grandpa and the War Department. . . . I used to get the mail early in the morning and whenever there was a long, businesslike envelope in the stack, I would know that Grandpa

had got an answer from the War Department and I would run upstairs with it. Grandpa would lift his head from the pillow, take the letter from me and open it himself. He would stare at the black print for a long time, then reluctantly, distrustfully hand the letter to me. "Well?" he would say. And I would read him the letter—reading slowly and pronouncing each word with extreme care—telling him that his claims for a pension had not been substantiated and that his application had been rejected.[1]

What makes this episode of interest is, first of all, the dates. *Black Boy* was published in 1945—a moment when information technologies of documentary identification had widely, but only recently, stabilized. The distance between Wright's book in 1945 (just as a new set of benefits were being distributed to a new generation of soldiers via the 1944 GI Bill) and Wright's experiences as a boy in 1920 or 1921 would have been noticeable. But even more glaring would have been the bright line separating Wright's retrospective glance in 1945 and the experiences of his grandfather over the half century stretching back to 1865. What is of interest is precisely that historical distance separating a pension system designed in the middle of the nineteenth century from its retrospective survey in the middle of the twentieth.

Civil War pension programs were in many ways the first iteration of the American welfare state, but their rationale and administration was enormously different from later Depression-era programs that continue to form the core architecture of state welfare provision.[2] In part because of such differences, what Richard Wilson had to accept in the late nineteenth and early twentieth century soon came to appear to many of his grandson's readers in 1945 as an acute injustice.

The impact of simple-seeming errors in the paperwork of "human bookkeeping" was increasingly noticed in the 1920s and 1930s as emergent administrative scaffolding expanded through bureaucracies and businesses. As an increasing anxiousness around such bookkeeping began to take hold, such errors as that of a misprinted name could come to be regarded not only as burdens but also as problems needing remedies. When remedies were developed (e.g., in the form of more standardized information technologies), the troubles that veterans like Wilson had suffered for more than fifty years could come to be recognized as what they effectively had been—namely, injustices. But such injustices were not always redressed by new information technologies. In some cases, newer technologies just made older injustices feel more inevitable than ever.

Wright's episode offers a vivid depiction of how information technologies of identity can facilitate injustices, inequalities, and unfreedoms. As information technologies of check box–clad printed blanks, filing systems, processing protocols, and computational analytics played an increasing role in American life from the mid-1910s to the mid-1930s, they began to assume the functional exercise of what I have been calling *infopower*. We today are not only brave new citizens confronting a newborn world of digitized data—we are also legacies of a power of information that has been constructing our subjectivity for almost a century now.

The Power of Information: Techniques, Operations, and Subjects

Consider a few contemporary tendencies that have recently become commonplace: online social profile management, the precision demographic categorizations of marketing analytics, unmatched levels of state-sponsored surveillance, the deceptively familiar cyberwars conducted by governments globally, powerful technologies of consumer and voter dataveillance, unprecedented levels of voluntary personal information sharing via social media, massive torrents of online file sharing, algorithmic and automated market transactions, data-driven financialization, and site after site for the production of informational aspects of ourselves from personalized genetic reporting to the quantified selves of our wristwatches. Though this is a seemingly disparate congeries of data points, there is one thing running through it: the fine thread of the power of information.

We were already spooling the thread of data a century ago when we initialized universal formats for persons on humble forms, plain cards, ordinary dossiers, and unassuming documents. This work of formatting is now—and has been for one hundred years—a political work. It is an exercise of power. It makes possible some of the ubiquitous features of our political landscape: the undocumented person, the young student tracked because of a risk for neuroticism, the victim of a gesture of racial discrimination made by no one in particular because it is the effect of a program querying a database. The burdens and benefits laid upon us as subjects of information rely on a work of formatting that conditions their very possibility. The work of formatting through which we become our information is political because without those formats, the particular burdens and benefits that rely on them could not take the same form. Information's formatting is a work that prepares us to be the kinds of persons who not only can suffer these inequalities and unfreedoms, but can also eagerly inflict them, often unwittingly, on others who have also been so formatted. Information thus became political precisely when we became our information.

If this is right—and perhaps even if it is only possibly right—then we find ourselves amid a swarm whose forms slip through the holes in our inherited conceptual nets. We need to mobilize a richer conceptual array in order to critically engage with contemporary infopower. We need to, in other words, transform the work of critical philosophy, rather than algorithmically applying old philosophy to new subjects. Every new generation facing new political, epistemological, and ethical conditions is in need of new philosophical bearings.

DATABASING INFOPOWER

Consider the possibility of scripting into a database the histories of the informational accoutrements I surveyed in this book's preceding chapters. Such a database would be populated with records demonstrating the emergence of a politics of information in the first decades of the twentieth century. Such a politics would be most visible in the collection of formats that this database would warehouse. Consider, for instance, how many such formats would be found in just the single case of early twentieth-century birth registration: formats of data solicitation of numerous types (date of birth, name of baby, name of parents, sex, race, parental occupation, legitimacy, etc.), formats established by registration procedures (dictating how agencies are to handle data they receive), formats established by information storage protocols (dictating how to save and retrieve information received by administrators), formats established by information-sharing policies and laws (dictating conditions under which data can be disseminated—for instance, to preserve privacy rights), and formats necessary for conducting an auditing of the entire process (dictating a shape to many of the data types implied by the formats above so that they are susceptible to a numeric, regular, and accurate accounting). This quick list may already seem excessive, but it is also far from exhaustive.

The database I have imagined could never be completed. Not only do we incessantly produce new formats that specify conditions for the manufacture, processing, and distribution of information, but every format generates new meta-formats that specify how to handle these formats as themselves data. This is why it is not an overstatement to say that formats are a torrent rushing in on us. Such a torrent is a danger—in part because it may drown us in a data deluge, but more so because in the rush itself is a subtle exercise of power that fastens what we can do and who we can be to particular formats.

The unprogrammable database of the last three chapters offers an array of instances of infopolitical fastening. From that array we can distill a general pattern of infopower at work. According to that pattern, infopower is enacted wherever we are formatted. Formats are acts of power that subject us to operations of being fastened to data. Again, we are fastened in a double sense:

pinned down and sped up, boxed and quickened, canalized and accelerated. We are first wound up tightly, and then set to a dizzyingly fast spin.

The specifics of infopower's implementations are as equally important as its general pattern. To limn more finely the contours of infopower, I offer a hypothetical query run on my hypothetical database to find three representative instances of the fastening performed by formats, one for each of information's functions of input, process, and output. Such a query, as I would code it, yields not examples of an independent theory of infopower, but rather cases (or records, or data) out of which a concept of infopower would be formed.

INPUTS. Consider the work of fastening in the context of data collection techniques for such monumental tasks as inaugurating a standardized birth registration system or enacting regularized operations for social welfare accounting in the first decades of the twentieth century. How was what we have since come to call "personal information" collected at its inception? In considering this question, it would be a mistake to assume that the personal information in question was already lying around waiting to be collected. Such an assumption would cause us to miss entirely the fastening operations of formats as they function in data input routines, for the formats themselves brought the personal information of informational persons into being. Without the formats that made operable such input forms as the Social Security Board's SS-5, there would be no possibility of assigning a Social Security number. This may sound like a strong claim, but I mean it as rather something like a banal truism. To grasp its truth, all you have to do is to imagine what would happen to an improperly completed SS-5; or, if you prefer to keep the argument on the empirical plane, then you could submit an improperly completed SS-5 (the most recent revision being 08-2011) and find out what happens. A Social Security number will be assigned only when an applicant completes the SS-5 according to the formats established by the form and its correlative databases, protocols, administrative personnel, and further such forms. When that number is assigned, it will always be according to the formats first established by the Social Security Board. This truism is an effective reminder of what we already know about ourselves as informational persons and yet are frequently invited to overlook. We so often look past the work that formats perform in fastening us to our data. Yet we know that it is only once we are fastened that we will be able to coordinate so many of our actions (with one another, with bureaucracies, with our social media friends). In so much of what we do, conditions of interaction are established by formats on forms. These formats establish the personal information of informational persons.

PROCESSING. Consider now the work of fastening in the context of the analytical algorithms that formed the heart of psychometric research programs at their emergence. How did personality psychology harness the streams of data generated by its questionnaires and channel those streams toward the production of scientifically sound inferences about the personality traits of the people filling out those questionnaires? The technique of the format and its attendant operation of fastening are here made visible in the information processing at the core of these tests. For one of the most important innovations of these tests was their use of algorithms to perform a number of tasks: converting natural-language answers (in the form of binary yes-or-no replies or natural-language response scales ranging from "never" to "always") into quantitative data (in the form of numbers), averaging these quantities (on either a per-question basis or a per-question-group basis) across a population of respondents (assumed to be either a random sample or a representative normal sample), matching these averaged results to a predefined set of natural-language descriptors of the personality features they were taken to represent (either on the basis of additional psychometric instruments, further findings of the test's developers, or inferences taken as warranted), and coordinating all these procedures in sequence according to a meta-algorithm that served as the controlling program for the entire test. These tests exhibit a great variety in their specific algorithms, of course (and contemporary tests such as those employing factor-analytic methods are exceedingly more complex than the simplified model I have described). But a constant for all of them is their use of algorithms to format and reformat those data fed in by test questionnaires. I do not propose to judge these uses, but only to focus attention on the formatting work that these algorithms enact. This formatting work functions to fasten test takers. An example of this is the impossibility of being taken by these tests to be the bearer of two personality traits defined as opposites. This is impossible because the test algorithms are constructed on the basis of an assumption of normal (modal) distributions against which all test takers must be measured—these tests cannot compute a set of responses on the test as indicative of an aberrant (bimodal) distribution of personality traits. In Allport's *A-S Reaction Study*, for example, a test subject cannot be both ascendant and submissive at the same time. The only options allowed by the test's algorithms are those of being either ascendant, or submissive, or neither particularly ascendant nor submissive. These tests pin their subjects in innumerable ways before we ever put pen to paper. They fasten test takers to predefined categories and simultaneously make it possible to quickly assess large swaths of them (be they students, army recruits, job applicants, prison inmates, or clinical patients).

OUTPUTS. Consider finally the work of fastening in the context of data dissemination techniques leveraged by technologists of race in the emergent years of a robust racial informatics. These data of race were presented in such forms as austere standardized appraisal reports and visually rich redlining maps. Such outputs were generated on the basis of established information-intake procedures and data-analytics processes. Those intake forms and processing programs were as much an innovation in the context of real estate racialization as they were elsewhere, and with a political effect that certainly appears more prominent. But their particular prominence is above all a function of the eye-catching condensations performed by their data output program. Redlining maps not only command our attention as evidence of racist policy at a state agency charged with regulating the most valuable financial asset of most Americans—they also draw our eye. They present data in a way that has all the appearances of the kind of quantitative accounting we regard as anything but political. Yet these maps are a politics unto themselves. Besides facilitating decades of discrimination, they are themselves built on the basis of race having already been formatted into data. Redlining maps are beautiful and tragic reminders of how information has become a near-transparent film that can be laid on top of anything and everything—including human skin, whose many shades thereby become discrete data points.

Formatting in each of its stages of inputting, processing, and outputting instantiates a power that conducts our conduct. It prepares persons for untold many uses, and also unexpectedly many abuses. This work of formatting is not itself necessarily oppressive, dominating, or otherwise forceful. Formats dispose rather than coerce, to adapt a compelling formulation from Davide Panagia's discussion of the mediatic bearings of Foucault's concept of the *dispositif* (a term that resists translation but resonates with my use of *assembly*).[3] Functioning as dispositive, formatting can both limit our freedoms or open up our liberties. Formats thus form a treacherous terrain: a field of power. This is why we need to multiply our means of attending to their dangers as they stalk us through their more obvious opportunities.

AN ANALYTIC OF POWER

In the introduction, I offered a schematic presentation of the model of power developed in this book: "infopower" is a distinctive *mode* of power; its "formatting" refers to a general array of *techniques* on which this power relies; its "fastening" refers to the way in which power is *operated* via these techniques, in a dual project of both "canalizing" and "accelerating" operations; and its "informational persons" are the *subjects*, or targets, of this mode of power.[4]

Beyond these three dimensions of power, it is also worthwhile to consider the *mode of rationality*, or *style of reasoning*, with which infopower affiliates.

Infopower is exercised in concern with a "data episteme." By this I mean to suggest that information itself is today a formidable mode of rationality by which entire domains of knowledge are made possible. The data episteme is an epistemology in which the need for more and more data is the spawn of data itself. Consider how frequently information is presented as a sufficient premise for other information. For nearly every bit of information we require someone to cough up, our need for that information is premised almost entirely on other information. We need more information because we are swaddled in so much information from the start. Think of all those forms you fill out just to fill them out, and with utter disregard for the contents with which you fill them out.

Consider now the countering worry of critics, past and present, that knowledge has been lost in our wildernesses of information. On my argument, this familiar worry does not register the loss of knowledge itself (as if knowledge could only ever be one thing), but rather indexes the changing terms by which knowledge is produced and exchanged. Data in many instances may not yet be knowledge, but so often today it is data that best makes knowledge. One historian of cybernetics registers this epistemology as one that demands "information, not authority, as the basis for action."[5]

Critics who remain unconvinced would do well to consider how they might set about disproving this point. There are, to be sure, countless ways in which you one could seek to dispute the claim that ours is an episteme of data. But surely the best way to build a disproof would be to whomp up a demonstration by way of data. The most convincing argument would look something like a survey of actual instances of knowledge claims, then analyzing them for the style of proof that is used to establish them, recording the results in a database, computing the relative frequencies of the different kinds of proof, and then designing a visual display of your results. Your bar chart at the end is what would seal your argument, even if the whole affair would be grounded in a speculative normative hypothesis you might muster right now before you collect your data. My point is only this: your hypothesis without your data is bound to be far less persuasive than your hypothesis demonstrated by data. Even the critics of data find themselves helplessly drawn in to its use. That is data as episteme.

Having excavated in detail the functioning of infopower above, I want to extend my schematic categories to a more general analytical model for power in a wider multiplicity of forms. According to this analysis, power can be analyzed into (at least) four constituent categories: *technique, operation, subject*, and *rationality*. A *technique* is how a mode of power operates. An *operation* is what a mode of power does. Techniques and operations are mutually constitutive: every technique of power exercises specific operations of power (i.e., there are no empty techniques that, as it were, do nothing), and every operation of power is exercised by techniques of power (i.e., there is no power operative in the abstract without instantiation in particular apparatus).[6] The technical operations of power are also always trained on specific kinds of material that I refer to as power's target or *subject*. The role of these technical operations in the formation of who we are and what we can do should not be underestimated. This is why, throughout this book, I sought to analyze the power of information in terms of technical minutiae of formatting rather than as the result of the grander forces of more canonical and classical political theory. One way to state this importance is in terms of the idea that technical operations and the subjects they target are coproductive and frequently enter into amplificatory reciprocal feedback.[7] This work of mutual adjustment between forms of subjectivity and power's technical operations is often mediated by a *mode of rationality* that works to explicitate these adjustment operations.

The generalizable analytic just sketched is only a heuristic device that I have employed to bring into view differentiable components of an assembly here under survey. This device proves helpful for distinguishing the concept of infopower from other conceptions of power that loom large in contemporary critical theory but which are unable to grasp the politics of information as such. Establishing these separations is crucial for the task of mobilizing the work of critique into new fields where intervention is needed.

The Irreducibility of Infopower: Beyond Biopower, Anatomopower, and Sovereign Power

Perhaps we need to do to Foucault what he did to all political theory before him: learn from him what we can and then productively press onward. One way to do this would be to open up what Foucault and Deleuze once called our "toolbox" in order to fashion new critical instruments and create new philosophical concepts.[8] In the context of an argument about the presence of infopower in the present, what does this involve? It does not imply that we should dismiss sovereign power, biopower, and disciplinary power as inappropriate to the contemporary. Nor does it imply that we need to demonstrate that infopower comprehends the totality of our narrow now. The guiding idea is rather one of enriching the conceptual repertoire of critical political theory.

This effort can be framed through the media archaeologies of Friedrich Kittler, whose work argues that, just as Foucault shifted the attention of political philosophy from the operation of laws to those of norms, we need to now shift attention once again—this time from norms to standards.[9] My focus on the formatting work of infopower is parallel to Kittler's attention to standards. But it is not identical with it. For one thing, Kittler focuses on standards as applicable to all technical media in general, whereas my effort is to isolate techniques of formatting enacted by specifically information technologies. For another, I am focused on the functional work performed by formats such that standardization is only one way to achieve technical functionality with a format.[10] These minor differences noted, Kittler's trichotomy helps stage my argument.

On Kittler's view, laws, norms, and standards are three different species of rule sets guiding human action. His emphasis is on their difference. Each is irreducible to the others. Foucault has undoubtedly shown that norms are irreducible to laws. Kittler conceptualizes standards as irreducible to both norms and laws.

It takes some consideration to see how Kittler's standards are not a return of the law. Standards seem similar to laws in that both operate in a binary fashion: you either abide by them or not. But the crucial difference is that laws produce a binary exclusion, whereas standards refuse to prohibit. Sovereign law divides the permitted and the forbidden. Law is deductive and negative in its operation. The problem with a political philosophy focused entirely on the law, as we learn from Foucault and Deleuze, is that it leaves no room for operations of power that are productive and generative. Standards do not share this feature of laws. A standard forbids nothing. It only specifies what you must do in order that you may do something else. Formats do not negate, but rather tether and tie our actions to specific shapes on the condition that we may want to assume these shapes for other purposes.

Kittler's standards (like my formats) thus share with Foucault's norms (both his normalization and his normation, for those keeping score[11]) the quality of enacting rule not by taking away but by eliciting, inviting, and producing. Sharing that much, however, Kittler distinguishes his standards from norms in that the latter "are an attempt to cling to natural constants," whereas standards are more "intentional."[12] One can see Kittler's point here, but it is a misleading way to put the matter. The more important difference between standards and norms is that the latter encourage conformity to averages (signaled by Foucault's double use of *norm*, connoting both what is socially encouraged but also what is statistically average along a normal curve), whereas the former invite, and create the very possibility for, adherence to new specifications. Formats (including standard formats) do not produce with an eye to averaging, but rather produce with an eye to specified designs to which we become fixed.

Kittler's political theory of standards brings into first view basic differences between my formats and the laws and norms of received political theory. But the argument needs more detail than Kittler's broad strokes offer. To fill in such detail, I shall deploy the analytic of power sketched above. This analytic offers a means for more finely differentiating the techniques, operations, subjects, and rationalities of infopower from the correlating elements in laws and norms. If such an analysis can show infopower to be sufficiently distinct from other modes of power, then infopower will have been shown to be irreducible to these other modes of power. Irreducibility does not entail incompatibility. Data composing informational persons can be leveraged for disciplinary normalization, biopolitical regulation, and any number of other such purposes. My argument, then, is just that there is something unique about the fastening performed by infopolitical formats that cannot be wholly comprehended by the concepts that speak to these subsequent uses.

I need to take care to differentiate infopower in detail for at least two reasons. The first is that the analyses of the previous chapters may seem initially to be comprehensible in terms of one of the three inherited models of power familiar from Foucault's work. Chapter 1, in its concern with the state and its legal mechanisms of birth registration, may appear to be an episode in the history of sovereign power. Chapter 2, in its focus on psychology as productive of distinctive individualization, may appear to be an account of the operations of disciplinary power. Chapter 3, in its focus on the workings of racism within gargantuan projects of social welfare, may seem to be comprehensible in terms of the mechanisms of social defense typical of biopower. Since infopower is not incompatible with these other modalities of power, I of course accept that each case offers evidence of operations of these other modalities of power.[13] My argument, however, is that comprehending the politics at work in each case has required extending our analytical attention beyond the limits of sovereign power, disciplinary power, and biopower. My claim is not that infopower replaces or negates these other modes of power—only that it operates beyond their limits.

A further reason for detailing the irreducibility of infopower concerns the incipient, and yet underdeveloped, role of information technologies in Foucault's own analyses of discipline and biopower.[14] Again, I accept that an inchoate tactics of information can be located within these predecessor exercises of power. My argument is that these tactics at some point began to assume a political gravity in their own right as they were increasingly employed to format and fasten our subjectivity. The power of information can thus be seen to be layered on other modalities of power. The layering of power suggests not negation, but addition and stratification. If multiple modalities of power are thus operative in the cases I have analyzed, that does not mean that their weighting is therefore equal. In contexts where the power of information assumes a greater gravity, its theorization becomes an urgent task for contemporary critical theory.

BIOPOWER

Foucault's concept of biopower is the most likely candidate within which my concept of infopolitics might be thought to be nested.[15] The very subject matter of my analysis of the infopolitics of racial data presents a challenge to my argument about the postbiopolitical status of data regimes, for Foucault himself clearly linked modern racism with the operations of biopolitics.[16] This seems to have already anticipated the infopolitics of race, which on my analysis made use of statisticalization to present analyses of racialized populations. So how is the informatics of race not just an instance of biopolitics?

Consider that biopolitics is and must be a politics of life. Its *subjects* are living beings construed as populations, or non-individuated sets.[17] Its *operation* is that of regulation, typified by such *techniques* as public health policy, demographic management, and medical intervention.[18] As described in *The Will to Know*: "supervision was effected through an entire series of interventions and *regulatory controls: a bio-politics of the population.*"[19]

Numerous theorists have employed Foucault's concept of biopolitics as a touchstone for the analysis of contemporary cultures of data, media, and networks.[20] On my argument, biopolitics contrasts strongly—and not just superficially—with these contemporary forms. Further, it also contrasts from their historical precipitants.

Consider my analysis of the informatics of race. This informatics apparatus certainly involves work that can be leveraged for the statistical treatment of populations, but it is rarely about populations taken as specifically *living* beings; nor is it necessarily about the *regulation* of these populations. The informatics of race is concerned less with the politics of life, and more with the politics of accounting. The accounting technologies I surveyed did not work so much to regulate the process of appraisal as to format appraisal technologies. Such appraisals of course are always usable as the basis for subsequent regulatory policies. But that a political technology can be subsequently leveraged for other purposes does not show that this technology must be wholly identified with those purposes. Rather, in such cases, the leveraging (here, biopolitical regulations) expresses those other purposes, rather than their being directly expressed by the technologies (here, an informatics of race) that are the subject of that leverage.

Of course, infopolitics and biopolitics can and sometimes do overlap. This is clearest with respect to techniques of statistical analysis.[21] Though statistics is quite prominent in biopolitics, there are nonetheless uses of statistics that are not strictly biopolitical in function because they do not operate the work of regulation over living populations. The selfsame technique can be operationalized for different purposes in different contexts.[22] If that is right, there is no basis for generalizing all uses of statistical reasoning to biopolitics.[23]

Nor is there a good reason to generalize all uses of information as power to the sole style of statistics. There are aspects of the informatics of race that are irreducible to statistics. The informatics of race was never purely statistical in the data assemblies it mounted. In the history I have tracked, the statistical processing of racial data was always contingent on a preliminary datafication of race that was not statistical, but rather a technical act of formatting. There is a politics of racialization at the level of printed blank forms with their little check boxes. These forms and boxes coded histories of phenotypical difference into differential data categories. They enacted a politics that fastened people to their data. Persons who were previously racialized by a different set of techniques began to become fastened to their race through informational technologies of forms.

For us today, one hundred years later, it is now perfectly familiar to engage one another as racialized on the basis of nothing other than the records of such forms that may be warehoused in a company database that a civil service employee references in the course of a phone conversation with a client. The politics of race, for so long reliant on phenotype, was coded so as to be seamlessly transmittable telephonically and digitally. Thus does race persist on the internet, despite early techno-utopian promises to the contrary.[24] Thus does it also persist through informatics techniques that no longer need to prompt people to tell them their race where data analytics can simply (and accurately) infer a person's race.[25]

ANATOMOPOWER (OR DISCIPLINE)

Anatomopower targets its *subjects* at the corporeal level.[26] As discipline is described in *The Will to Know*, it is "centered on the body as a machine: its disciplining, the optimization of its capabilities, the extortion of its forces, the parallel increase of its usefulness and its docility, its integration into systems of efficient and economic controls, all this was ensured by the procedures of power that characterized the *disciplines*: an *anatomo-politics of the human body*".[27] As fully detailed in *Discipline and Punish*, discipline *operates* a power of normalization by coaxing bodies (not physically coercing them) to conform to the norm.[28] Its *techniques* include panoptic surveillance, regular examination, and a meticulous training, or dressage.[29]

Disciplinary tactics—and perhaps most notably panopticism—have been a theme of much interest for contemporary thought on the politics of data, and in particular analyses of digitized surveillance and dataveillance.[30] This work shows that disciplinary anatomopolitics, like biopolitical regulation, frequently establishes productive affiliations with something like what I have been calling infopolitics. But, again, this does not make the latter reducible to the former.

First, if discipline relies on such mechanical techniques as an overt surveillance that announces its own visibility, the work of infopolitics is often dependent on informatics operations that resist being seen.[31] In the case of the personality profiles analyzed above, the algorithmic processing constitutive of the tests is hidden beneath the paper questionnaires visible to test takers. Indeed it is a general feature of psychometric testing that its computational back end be a black box to test takers (otherwise subjects could, and rather well would, game the tests).

Second, infopolitics differs from discipline with respect to the normative or prescriptive operations of disciplinary power. There is, as Foucault's *Abnormal* lectures clearly state, always a normative dimension to disciplinary normalization: "The norm's function . . . is always linked to a positive technique of intervention and transformation, to a sort of normative project."[32] By contrast, infopolitical formatting does not normalize us—it does not tell us how we should act or how we must act. The work of fastening is focused less on norms and more on forms. The forms themselves are not trained upon what we should become, but are focused on delineating the shapes that we already are. Forms enact an informatics of fastening that dispose and bind us to their specific formats. There is no conceit of coaxing us to conform to these categorizations (though discipline may seek to do that to us later). Rather, these categorizations name features (like traits) that we already possess, or which are already in possession of us.

Third, anatomopolitics is targeted at corporeal materiality in a way that infopolitics often entirely dispenses with. Personality profiles do not pin us to our bodies, which are only incidental appendages of the apparatus requisite to fill out the questionnaires. They pin us rather to our traits, which are construed as anything but physical and material. Traits are correlates of information technologies like the test. The personalities that these traits compose are, for better or worse, about as disembodied as a thing can get. That is why you can take a personality test online. It is also why bots can take the tests and be reliably assigned personality traits. It is why computer programs can have personalities and why personalities can be uploaded into computers.

SOVEREIGN POWER

If the politics of information demands a conceptualization beyond the familiar limits of biopolitics and anatomopolitics, then it might be thought that the politics of information works to revive power in "its ancient and absolute form" of sovereignty.[33] As analyzed by a venerable tradition in political theory paradigmatically consolidated by Max Weber's definition of the state as a monopoly on physical violence, the power of sovereignty is a mode of power operating through force and brutality.[34] Foucault's work suggests that sovereign power often assumes subtle variations, and in so doing should be thought of as expressive of "power in a binary system: licit and illicit, permitted and forbidden."[35] This is the sovereign operation of prohibition effected through techniques like law.[36]

In functioning in lawlike fashion to forbid and permit, sovereign power might appear to be a mirror for the command structure of information systems, especially those contemporary information systems that seem to simply tell us what to do. This is the parallel that legal theorist Lawrence Lessig brings into view with his famous claim that "code is law."[37] Lessig's argument, however, depends on an analysis of information systems as always functioning deterministically rather than probabilistically. But, as Wendy Chun shows, code never simply does what it says.[38] Nothing is self-executing: not code, not even the law.

Complicating the assumptions about power that a claim like Lessig's must rely on, Tung-Hui Hu has more recently argued that contemporary data culture "indexes a *reemergence* of sovereign power" now "under different forms."[39] Hu's persuasive claim is that "the cloud grafts control onto an older structure of sovereign power, much as fiber-optic networks are layered or grafted onto older networks."[40] Hu's compelling appellation for this "graft" is the "sovereignty of data."[41] I want to not so much disagree with Hu's insightful analysis as to complicate it in turn.

Hu's concept of the sovereignty of data appears applicable to numerous contemporary information technologies, but I find this concept difficult to apply to the specific histories of the information technologies I discussed in preceding chapters, and in particular the technologies of registration and enumeration in my analysis of state-produced identity documentation. This cannot be a merely incidental site of inapplicability. If one were to go looking for sovereign exercises of power that have been exerted over informational persons in their historical production, then surely one of the first places one would look would be at those state agencies that set up binary systems of inclusion and exclusion on the basis of databases of enumeration and registration. For example, one might think of the expulsive work of immigration and border control that depends on technologies of birth registration so as to exclude (many of) those who are unregistered. Yet such sovereign acts of inclusion and exclusion were subsequent operations that leveraged previously consolidated databases of births, workers, and so much else.[42] Such sovereign acts may produce prohibitions.[43] But the databases themselves do not prohibit, forbid, or exclude anything. They only format that which would become warehoused as data. This difference is crucial to understanding how infopower is neither already sovereign power nor always grafted to it.

Formats, just like Kittler's standards,[44] do not function in sovereign form insofar as there is nothing in a format that forbids anything at all. A sovereign garishly expunges, or extinguishes, that which is deemed impermissible. But a format leaves everything as it was such that a subject that is not formatted according to its terms is not committed to nonexistence but only consigned to either try again or go its own way.

If the power of information is at all expressive of sovereign power, then it can only be expressive of a decidedly altered shape of sovereignty. It would be a power that dispenses entirely with techniques of violence and logics of prohibition and permission. Hu considers precisely such a transformative redescription in referring to the "sovereignty of data" as a "graft" of sovereign power and technological control, or what he at one point calls a "hybrid construction."[45] But Hu emphasizes the continuity inherent in such an amalgamation, placing the weight of his argument on "a *reemergence* of sovereign power within the realm of data."[46] By contrast, I emphasize the manner in which contemporary data power introduces difference with respect to historically preceding modalities of power. My preferred metaphors are thus rooted less in the continuous concepts of botany and more in the discontinuous concepts of geology. Infopower is deposited on, or layered on, the sediment of earlier strata of power.[47]

OTHER POWERS OF DATA

Numerous other theorists of the politics of data have adopted approaches proximate to my own strategy here of arguing for a fresh conceptualization of the power of information. The most prominent of these offerings build on Gilles Deleuze's famously obscure suggestions about a new "control power."[48]

Alexander Galloway argues that we face today a new protocological control; Maurizio Lazzarato develops a novel idea of noopolitics drawing on Deleuzean control; and Rita Raley refers to Deleuze's analysis in her influential argument about tactical media.[49] Other theorists have made productive use of Deleuze's concept of control alongside Foucault's analysis of biopower, such as Tiziana Terranova's argument about "communication biopower," Wendy Chun's work drawing on Deleuzean control and Foucauldian governmentality, and John Cheney-Lippold's conceptualization of a "soft biopolitics."[50] Beyond appropriations of the famed conceptual architectures of Foucault and Deleuze, a number of alternative theories of data power have been recently forwarded. Grégoire Chamayou presents "datapower" as a "technology of power" that is "less an apparatus of surveillance than one of control," though without aligning control to Deleuze's concept.[51] Bernard Stiegler has developed analyses of the history of our informational present in terms of a "psycho-socio-power" and a "psycho-power."[52] David Beer argues that data governs us by a "metric power."[53] Davide Panagia has explored the dynamics of what he calls "#datapolitik."[54] Louise Amoore theorizes a data-focused "politics of possibility," from which I drew in this book's introduction.[55] Closest to my own approach, and so deserving detailed discussion below, Bernard Harcourt has provided a Foucauldian theorization of the "expository power" of "the digital age."[56]

A crucial point on which my analysis needs be separated from all these accounts concerns the avant-gardism pervasive across recent theorizations of the powers of data. We are in almost every instance invited to regard new media as new, and by extension to regard their politics as taking new forms too. Such avant-gardism in the face of prevalent social media, state surveillance, and other universalizing information technologies is understandable. But it is also historically, which is to say empirically, unsupportable. An analysis of the power of information as having only recent relevance fails to confront the scale at which we have been invested by information for more than a century. The political dynamics that many recent critical analyses are speaking to happen to be much more deeply entrenched than their analytical concepts would suggest. Consider control and exposition, to take the work most proximate to my own analysis of infopower.

With respect to control power, the tendency to disregard the past in favor of the future is captured in Chun's apt observation that "Deleuze's reading of control societies is persuasive, although arguably paranoid," in the way that it "overestimate[s] the power of control systems."[57] This overestimation follows from a tendency to look not into the past with the empirical rigor of genealogy, but rather toward the future in a mode of eschatological prediction.

Deleuze's musings on control are rife with an anxiety about "approaching forms of ceaseless control."[58] Following Deleuze's presentation of himself as writing about "control mechanisms as their age dawns," new media theorists of control have similarly given in to such avant-gardism.[59] This is perhaps clearest in Galloway's arguments concerning what he calls, on the first page of his book, "a new apparatus of control that has achieved importance at the start of the new millennium."[60] Such avant-gardism can be sustained only by neglecting the past in favor of obsessing over the future, which is to say it can be sustained only by a style of critique that finds itself helplessly attracted to the potentiality of a speculative exercise that can never be historically checked.[61]

A less extreme avant-gardism is at work in that theory of information power that is both methodologically and conceptually most proximate to my own—Harcourt's theorization of expository power. What I find most insightful about Harcourt's analysis, as I noted in my introduction, is his attention to the shaping of subjectivity: "Our selves and our subjectivity . . . are themselves molded by the recommender algorithms and targeted suggestions from retailers, the special offers from advertisers, the unsolicited talking points from political parties and candidates"; and later, "Digital exposure is restructuring the self."[62] For all the insight afforded by his analysis, however, I dissent on a crucial point: my argument is focused on data technology, whereas Harcourt's attends to digital technology. These differing aims of analysis belie a crucial historiographical difference concerning the extent to which critical philosophy should bring the material of the past into view as a means of putting pressure on the present. Harcourt's analysis declines to examine how the digital relies on the informational as its historical condition of possibility. And this—information as a historical condition of possibility of contemporary shapes of subjectivity—is precisely what I have sought to attend to. If my analysis is focused on the emergence of a power of information when it was on the cusp of a consolidation, Harcourt is rather differently focused on characterizing a "new digital age,"[63] beset by "a new way for power to circulate throughout society."[64] This historiographical difference results in a divergence over how power grips subjectivity. Harcourt's conclusion is that expository power operates over our digital "doppelgänger" or our "data double" such that "the algorithmic data-mining quest of the digital age is to find our perfect double, our hidden twin."[65] Despite apparent proximity, this view contrasts strikingly with my argument that infopower targets us as persons actually composed of our data. My argument is that data has become a crucial part of the very terms by which we can conduct ourselves. We are our data. Therefore we are precisely not doubled by it.

Ultimately, my claim is that what is otherwise so insightful in recent arguments about expository power and control power needs to be properly embedded in a wider history. That which feels so recent in digital computation or in new media needs to be properly situated as a rippling on the surface of a vast sea of data subjectivity. This points back to the broader perspective of my argument. I have sought to bring into focus a series of historical operations from an analog past that continue to condition our digital present. What matters is not that the present is "new" nor that the past is "old," but rather that the consistency of past practices continue to operate in present practices.[66] It is not novelty that matters for political theory, but rather differences in how power is exerted and trained to produce differences in modes of subjectivity.

The Historical Layering of Power: The Craft of Infopower

Having differentiated my analysis of infopower from other prominent conceptions of power that are being employed to make sense of our data assemblies, I need to underscore that the point of highlighting these differences is not to undermine the theoretical tools made available by these other contributions to political theory. The point is rather to further the work of political theory by proposing another such tool. This requires showing why the proposed new tool is irreducible to those already furnished. Infopower is thus offered as a distinctive, but not exhaustive, instrumentality. A plurality of conceptions of power are always potentially relevant. And their relevance is always a question that requires particularized genealogical (or anthropological, or otherwise empirical) inquiry.

Such a pluralism about power raises important questions concerning the historical relationships between differing modes of power. How do different modes of power relate? To respond to this question, I employ a metaphor of layering.

Infopower constitutes a layer or stratum in the recent history of power, not a reigning new successor that displaces biopower, discipline, and sovereignty, nor a new social totality that covers them over. But what is at stake in describing power in terms of regional layers of practices, rather than temporal eras or social totalities? The question of the relation between different modes of power is a crucial one for genealogical political philosophy that has generated much recent discussion.[67] Prompting these debates is Foucault's lack of clarity about the relations that hold between divergent operations of power.[68] Nonetheless, Foucault clearly develops one idea in a few key places, including this prescient passage from his 1978 Collège de France lectures: "There is not the legal age, the disciplinary age, and then the age of [biopolitical] security. . . .

We should not see things as the replacement of a society of sovereignty by a society of discipline, and then of a society of discipline by a society, say, of [biopolitical] government."[69]

Verena Erlenbusch-Anderson captures the crux of Foucault's approach: "Resisting notions of replacement and obliteration, Foucault emphasized interpenetration and superimposition."[70] In Erlenbusch-Anderson's genealogical analysis of terrorism, this insight proves valuable as a guide for tracking techniques of terror as they dip in and out of differing regimes of power. For my project, this insight helps track information's sometimes-simultaneous role as a technical implementation of an emergent modality of infopower that can be layered on, for example, biopower and discipline. Following Erlenbusch-Anderson's lead through Foucault, then, my argument is that the historical relationship between infopower and other of power's modalities is neither negative nor substitutive, but rather additive, or layered.

This historical relationship can be summarized as follows: a craft of information was born within biopolitical and anatomopolitical contexts and from there slowly invaded multiple domains of each, as well as a number of institutions fundamentally oriented by sovereignty. In its earliest appearance, information was itself a mere instrument of other modes of power. Later, that craft of information assumed a gravity of its own such that it became a distinctive assembly of power in its own right: an infopower of fastening, with its own tactical operations, meticulous techniques, enlisted subjects, and correlative rationality of data.

We downplay the power of information too much if we see it as merely an appendage of modes of power whose functions are essentially targeted at populations, bodies, and territories (and so on) through techniques of regulation, normalization, and permission (and so on). Infopower is a mode of power in its own right in that it operates its own distinctive set of techniques that exercise distinctive operations of power on distinctive kinds of subjects. Infopower makes use of techniques of formatting in order to operate a fastening, canalizing, and accelerating that is targeted toward the many kinds of users, accounts, and records that we informational persons have become.

Data's Turbulent Pasts and Future Paths

When Communication Is Not the Problem

Philip K. Dick's 1955 short story "Autofac" poignantly captured the concern of its moment.[1] Situated in a postwartime future, a band of humans desperately attempts to establish communication with an automated factory apparatus. "The Institute of Applied Cybernetics" designed the autofac to produce everything that would be needed by the humans who would survive the devastating war. Their design was so perfect, their autofac so effective, that it eventually begins overproducing its goods, threatening the depletion of environmental resources.

Dick's story appeared in a heady moment. It was published only a few years after Norbert Wiener's celebrated *Cybernetics* of 1948, Claude Shannon's "A Mathematical Theory of Communication" of that same year, and Alan Turing's groundbreaking elaboration of artificial intelligence in his 1950 essay "Computing Machinery and Intelligence."[2] Turing famously argued that a machine could be said to exhibit intelligence if a human tester could not discern the difference between information outputs sent from the machine and the same sent from another human. The conditions for the Turing test involve physical isolation so that the tester only receives identically formatted printouts from both test subjects. The test thus strips from machine and man alike much of what we traditionally associate with each. Reduced to the barest behavioral signs, Turing asks, what distinguishes the communications of humans and computers?

This is the precise issue with which "Autofac" opens. "The problem we have here is one of communication," asserts one of the last humans.[3] He is attempting to establish communication with the autofac so that the humans can convince it to cease overproducing. Their first attempt is to destroy the goods sent by an autofac delivery drone. But the autofac only sends more. Then they tell the autofac that its goods are defective. But the autofac tests its products and certifies their condition. Finally, the humans establish contact by sending a meaningless message: "The product is thoroughly pizzled."[4] The term is "a semantic garble," one human explains to another.[5] In response, the central autofac sends a humanoid "data-collecting machine" that describes itself as "capable of communicating on an oral basis."[6]

In the ensuing conversation, it seems as if the autofac fails the Turing test when it admits of itself, in the third person, that, "although purposeful, it is not capable of conceptual thought; it can only reassemble material already available to it."[7] Yet the humans too fail, at a different test—convincing the machine that its products are defective, or even pizzled. For the autofac knows what the humans know all too well: its products are not flawless. The machine, it turns out, does not fail to communicate. It successfully communicates the truth. It just so happens that the humans do not want to recognize the truth. Written in a fevered cultural moment when communication was thought to be everything, Dick's story offers a glimpse of why communication is, at least sometimes, not enough.

In the story's conclusion, the humans break in to the central autofac and attempt to destroy it. But they discover that the autofac has been so effectively programmed that some of its parts can remake its other parts as quickly as they are destroyed. The autofac is a machine capable of making anything, including itself. The humans persist and eventually discover a section of the autofac buried deep underground. They discover there an apparatus producing small metal pellets and shooting them upward through a tube to the surface. Later, when they reemerge into daylight and locate a pellet, they find within it a microscopic machinery that is "working energetically, purposefully" to assemble an autofac clone.[8] Deep beneath the autofac's many surfaces—buried within its delivery machines, its humanoid ambassadors, and its every apparatus—is a robust data program that the humans never thought to interrogate. That informational kernel is precisely what any attempt to communicate could never address, for it would of necessity be the basis of any possible communication.

The Mobile Pasts of Data: An Alternative Historiography

Getting a grip on today's politics of information requires returning to the times when our "narrow Now," to borrow a crucial historiographical concept from W. E. B. Du Bois, was assembled.[9] In focusing on the narrow now of data, the history traversed in this book has reached back before the World Wide Web in 1994, back before the Apple desktop in 1976, and back before information theory's consolidation in 1948. I have examined an earlier and more turbulent moment in which information began to function as a universalizing technology for formatting and fastening. In that moment were composed resilient embankments through which our narrow now continues to flow. Those embankments were anything but inevitable. Our complicated canalizations could have been designed differently.

The alternative historiography I have adopted is meant to help bring into view the very idea that there is a politics in these designs. This book has argued thus far that we find ourselves today in the midst of a distinctive political problematic of information (as one problematic among many). It is crucial for this argument that this problematic not be mistaken for another familiar political problematic into which information is often read: that of the politics of communication. However tight the connection between information and communication is, it does not imply that a politics of information is already fully expressed in the politics of communication. This is a mistaken identification worth heading off, simply because it is all too inviting.

This chapter develops the argument that information politics is separate from communicative politics along two trajectories: one looking backward into the past and the other reaching forward into the future. I first examine in this section how a historiographical turn to information before information theory helps us avoid a continued reliance on a kind of communication-centric perspective that information theory itself helped consolidate. That earlier moment, because it was unconsolidated, lets us see information as political in ways other than that which a communicative perspective brings into view. A second argument, developed in the following section, considers the crucial question of normativity: How ought we to conduct ourselves in the face of infopower? Here I suggest that the seemingly most likely offering—the communicative perspective of deliberative democratic theory—is actually a nonstarter for interrogating the politics of information. This is an uneasy conclusion in light of the tremendous normative potential, and actual normative gains, of democratic deliberation.

THE HISTORIOGRAPHY OF THE
MILITARY-INFORMATIONAL COMPLEX

The genealogical analysis of informational persons and informational power I have developed offers an alternative to what I referred to in the introduction as the consensus historiography of information politics. That consensus focuses on a moment of cold consolidation and its frozen aftermaths. In focusing instead on a time preceding Cold War information theory and scientific cybernetics, I have sought out practices of information before information theory. My attention on the 1910s to the 1930s, rather than the 1940s and 1950s, brings into view techniques of informatics in operation prior to their later consolidation.

This earlier historical moment is of interest not because it was first or originary, but rather because it offers a view of the simultaneous presence of transforming mobilities and cementing stabilities. The preceding chapters thus attended to developing information technologies on the cusp of consolidation and transformable techniques on the verge of condensation.

Why do such moments matter? To get a sense of the stakes, consider how they form an alternative to one way in which the consensus historiography presents information theory. Media archaeologist Friedrich Kittler, whose attention to historical and technical specificities deeply informs my methodology, boldly asserts that "cybernetics, the theory of self-guidance and feedback loops, is a theory of the Second World War."[10] In Kittler's wake, numerous historians of science, communication, and information have developed their own versions of this militaristic origin story.[11] These arguments all assert that information theory originated in wartime exuberance for technologies like anti-aircraft ballistics and machine cryptography. This argument is not only common among contemporary critical theorists of information. It is also the precise origin story purveyed by the information theorists themselves. Kittler confidently quotes Wiener's assertion that "the deciding factor in this new step [toward cybernetics] was the war."[12] In light of the alternative history I have presented, we can now ask what gets lost when we allow the information theorists to narrate their own history in this way.

One thing that gets lost is the technical history of information theory itself— that is, the information technologies that helped produce information theory. Many of these technologies did not aim toward the military-informational complex that the consensus historiography would posit as their result. This is an elision that critics of Kittler have noted.[13] It is also an observation that a small handful of historians of the period have ventured.[14] What these alternative histories suggest is what we would expect of any serious investigation of the question: information theory congealed through a roaming multiplicity of causes. Accounting for this multiplicity requires, at the very least, attending to practices of information prior to their consolidation in information theory.

THE HISTORY OF INFORMATION
BEFORE INFORMATION THEORY

We need to free the critical history of information from a historiography that can too easily denounce information theory in light of its affinities for militarism, abstractionism, and other obvious errors.[15] In contrast to the posture of debunking, I am more sympathetic with the style of resistance expressed by Siegfried Zielinski in his 2002 *Archäologie der Medien* (translated as *Deep Time of the Media*, thereby losing the Foucauldian resonance): "A short in the cybernetic system—one cannot get the better of this programmed and standardized world by machine wrecking. . . . The only effective form of intervention in this world is to learn its laws of operation and try to undermine or overrun them."[16]

Rather than engaging in critical history to teach the lesson of why we should reject information theory—and presumably all the information it theorized—I have sought to contribute "a usable past," one that can direct our critical attention to political engagement on the transformable terrain of information itself.[17] For these purposes, I have no interest in denying the claim of the consensus historiography that there was a moment of consolidation. Rather, my approach involves excavating the conditions of that very process of stabilization.[18] Thus I have focused my analysis on what I have called a "cusp of consolidation"—a cusp presents us with transforming dynamics as they appear in their adjacency to the solid states into which they soon freeze.[19]

One way to observe mobilities as they slow down to an almost complete stop is to attend to controversies in the past that have since become stale. Alongside all the operators and boosters of information I have canvassed in this book, there were always eager critics of information itself. Among them was the man who remains very much the personification of the critic. In his 1934 "Choruses from *The Rock*," T. S. Eliot wryly asked:

> Where is the wisdom we have lost in knowledge?
> Where is the knowledge we have lost in information?[20]

Eliot's questions surely seemed poetic to many readers in 1934. For us today they are as cliché as one can get. Every schoolchild, swaddled in a sea of information that is barely knowledge and nowhere near wisdom, rehearses these questions at periodic intervals throughout the course of their education. But today's clichés may have been yesterday's insights.

Eliot's insight depended on an awareness of how stable the cultural technology of information had already become by 1934. One must be intensely aware of a swarm of information to worry about its possibly causing a loss of knowledge and wisdom. Eliot was perceiving a distinction between information and knowledge that was then beginning to settle into working order. That such a distinction was developing in the face of the period's informational abundance should be set against the backdrop of prior associations of informational content with epistemological achievement.

Consider the history of the concept of data. In "Data before the Fact," intellectual historian Daniel Rosenberg describes how in the eighteenth century, data "went from being reflexively associated with those things that are outside of any possible process of discovery to being the very paradigm of what one seeks through experiment and observation."[21] Prior to that shift, the concept of data had been associated with what was rhetorically given or assumed, but not necessarily proven. Then, in the eighteenth century, the heyday of classical empiricism, data became tethered to conceptions of discoverable fact and demonstrable knowledge. If Rosenberg shows that data in the eighteenth century shifted logical status from assumption to conclusion, then there was a re-reversal in the early decades of the twentieth century. Data were once again put to work as no longer in need of any epistemic backing. Data were becoming that which can be assumed or given. Thus could data remain perfectly functional despite their being recognized, already by 1911, as "a wilderness," or what we today call an "information overload."[22]

Eliot was worried about precisely that: an abundance of, and exuberance for, information devoid of epistemic status. His worries tracked the future with precision. By the mid-twentieth century, it was accepted that data no longer needed to be true—indeed, no longer even needed to be meaningful—to be data. This was the most decisive assumption of Shannon's information theory. Eliot's anxiety can retrospectively be seen as depending on an early twentieth-century reemergence of that pre-eighteenth-century split between knowledge and information.

If critics like Eliot sought to enforce the split so as to discredit data and save knowing, some of his contemporaries argued on behalf of the distinction with other interests in view. For pragmatist philosophers George Herbert Mead and Wilfrid Sellars, a new epistemology was needed to help move us past those very eighteenth-century theories that had taught that data was always given as knowledge.[23] In this, the pragmatists were surprisingly close to the information theorists whose arrival the critics forewarned. Even though the pragmatist epistemologists sounded like Eliot in that they cared most for knowledge, they were distant from Eliot in that they had no brief against data, so long as it was suitably contained and not confused for knowledge itself. Their conceptions could thus mingle happily with Shannon's, for his only concern was with data without regard to its possible truth or meaning. Such debates only indexed, and did not themselves determine, the transforming technical conditions of epistemology. As the poets and philosophers debated, the technicians quietly leveraged the reemergent distinction between knowledge and data so as to accept the technical ascendency of the latter. Indeed, information could be variously criticized (by its detractors) or consolidated (by its theorists) only because it had already been made real by tinkering technicians who were busily devising their many machines. What the critics were therefore correctly witnessing, even if only dimly, was that information had begun to breed right before their eyes. To employ a cybernetic anachronism, information had initiated a feedback loop, a self-perpetuating demand for more and more data without regard to its epistemic or semantic status. We still live within the riotous motions of that circularity: our information carnival.

This carnival was conducted not by its grand theorists, but by its quiet technicians. Thus I have sought to highlight in this book the work of unexpected actors who contributed designs, forms, and processes to the assembly of our contemporary informational milieu. The formatting of the American birth certificate was in part the work of bureaucrats at the Census Bureau, but its success also crucially depended on an all-volunteer network of women's clubs. The information technologies undergirding racialized redlining were not just the product of federal home finance agencies (and were barely the product of political representatives whose legislation established these New Deal agencies), but rather were ported to agencies from a network of real estate professionals who had sought more scientific practices of appraisal. The consolidation of psychometric science was in part the work of generals and scientists working on intelligence tests for the US Army during World War I, but it was also due in part to a small cadre of professionalizing psychologists who secured for their fledgling field of research the status of science.

Taking into account this broader array of actors helps unsettle predictable convictions about the intentions of those who would make us into our data. Behind big events there do not always lurk grand strategies like military campaigns, capitalist schemes, state power, and fantasies of social abstraction. Sometimes the technologies that eventually come to possess an enormous political and epistemic gravity are initiated in tiny designs by humble technicians. If this is right, then at least sometimes the intentions that created technologies matter less than the affordances unwittingly enacted through slender techniques. The designers of the standard birth certificate installed a series of formats that now helps define persons as accounts in manifold databases in which they are enrolled. The technicians of personality metrics struggled to define personality traits in terms making them amenable to forms of measurement that would soon come to define the futures of schoolchildren shown by these instruments to be neurotic or aggressive. The professional appraisers who redesigned the practices of real estate valuation built racial definitions into geography and thereby rebuilt racialized subjectivities on the basis of an informatics of race. These technicians did not harbor grand political ideals. They were modest makers of formats. Yet their formats afforded certain possibilities for power, knowledge, and subjectivity. They initiated the formatting of people who would live and act through their data designs. They precluded what could have been other options.

So what can we do to redesign these formats where they function today as programs for the disposition of burdens? And what forms of normative guidance can we offer for the ongoing development of next-generation information technologies whose humble technicians today may not fully appreciate the massive impact their work could come to have?

The Normativity of Infopower: Beyond Communicative Democracy

"I'm not a prophet; I'm not a programmer [*programmateur*]; I don't want to tell people what they should do. I'm not going to tell them, 'This is good for you, this is bad for you!' I try to analyze a situation in its various complexities, with its functions, for this task of analysis to permit at the same time refusal, and curiosity, and innovation."[24] Like Foucault, I do not aspire to become a *programmateur* (an organizer, a scheduler, a programmer) who would guide or code our times. The work of a critical diagnostics of our historical present cannot have as its only point the future to which it might lead. Like Foucault, I hold a problematizing conception of critique: "[Critique] isn't a stage in a programming [*programmation*]. It is a challenge directed to what is."[25] Critique is a clarifying practice, not a theoretical ensemble of directives.

In the wake of the critical diagnostics of data politics offered thus far, I turn now to the future-facing issues of the normativity of data orders. I do so not out of a desire for a definitive normative coding. My hopes are more for a preparatory assembly of designs that could further the work of decoding present practices.

Though my discussion of the normative dimensions of infopower will therefore be limited, it is also clearly occasioned by the genealogical diagnosis I have thus far developed. For if I am right that we occasionally find ourselves in the grasp of infopower today, then this diagnosis sets limits on what could be positioned as a normative resource for resistance to infopower's dangers. The major such limitation that needs to be explored—in large part because of its conceptual proximity to the domain of information itself—concerns normative democratic theories taking communication as their core. In his analysis of digital expository power, Harcourt argues that "the fact that we live in a democracy has facilitated, rather than hindered, our expository society."[26] If this is right, then my argument is that the tendency of twentieth-century political theory to locate democracy within communicative interaction may explain why the politics of information has remained, as Harcourt states, "largely *invisible* to democratic theory and practice."[27]

I develop this argument in two steps. First, I pick up from the historiography of information just discussed in order to clarify the historical relationship between information and communication. Second, I turn to communicative or deliberative democratic theory, considering both its most influential contemporary statement (the communication-theoretic account of Jürgen Habermas) and one of its chief antecedents that also happens to be from the period of survey that has been my focus in this book (namely, John Dewey's pragmatist theory of democratic publics in his 1927 *The Public and Its Problems*).[28] These two steps enable me to explicate an unexpected conceptual proximity between communicative theories of democracy and the assumptions informing the so-called "information theory" of Claude Shannon and Norbert Wiener. Excavating these proximities depends on my first showing that information theory, despite its familiar label, was in actuality a theory of communication that simply assumed information as its given starting point. This historical point suggests in turn a conclusion that should be consequential for communicative theories of democracy: if communication is the symptomatic response to an information fever, then the cause of that fever cannot be cured by its symptoms. Communication cannot address itself to information, but can only presuppose it. Every theory carries with it the limits of being unable to confront that which it presupposes. For all their advantages, then, the one thing that communicative theories of democracy cannot bring themselves to confront is the functional role of the politics of information within democratic polities.

INFORMATION THEORY IS ACTUALLY COMMUNICATION THEORY

In light of widespread misconceptions furthered by almost all contemporary scholarship on information theory, it is crucial to sort out the relationship between the two core concepts at work in Wiener's grandiose theory of cybernetics and Shannon's humble technical contribution to electrical engineering: the concepts of information and communication. Two points need discussion. The first concerns the importance of communication to a wide array of projects in the early decades of the twentieth century. The second point is that any project of communication must be dependent on prior projects of the formation of information. Without something to be communicated (a formatted datum that can of course always be reformatted, including by communication channels), there is no prompt motivating the work of communication.

The first of these points can be made by way of John Durham Peters's claim, in his definitive history of the idea of communication, that "all the intellectual options in communication theory since that time were already visible in the 1920s."[29] If Peters is right, then an obsession with communication preceded information theory by at least a few decades. Though it preceded the theory of information, this obsession did not precede information technology practices, such as those I have here investigated. Perhaps, then, the need for communication began to be felt by the very people who were beginning to understand themselves as information.

This brings me to my second point: that communication by itself cannot address information as such, but can only presuppose it. In one sense, this is a logical claim; any communicative act must always presuppose some information to be communicated. In a more robust sense, however, it is also a historical observation about the emergence of the logic of communication as having preeminent importance for the twentieth, and now twenty-first, century. My claim is that Shannon and Wiener in 1948 could mathematize and abstract information for the purposes of transmission, or communication, only because information had already achieved widespread use.

Shannon and Wiener both wrote with a confident assumption about the importance of information. In so doing they merely presupposed what almost all their contemporaries were already doing with data. The preceding chapters have shown that, at the time at which the information theorists were writing, it was already the case that for more than two decades the seemingly innocent information on any blank form could already be extracted from its context of origination for storage, processing, comparison, repurposing and distribution in any number of other contexts. What this suggests is that information in 1948 was already functioning according to Shannon's supposedly simplifying formalization: a signal to be transported without concern for its possible meaning(s).

By the 1940s information was sufficiently stable as a set of cultural practices and social technologies to serve as the basis for the assembly of a communicative culture. Put otherwise, information had become such a crucial matter of concern that entire fields of practice and inquiry could be developed in response to it. Consider, for instance, the communicative turn characteristic of mid-twentieth-century thought in a number of fields (such as the linguistic turn in philosophy). If this is right, then the information floods of the 1910s to 1930s were the practical conditions from which Wiener's and Shannon's theories began.

This argument also helps explain a little indisputable fact that much recent critical scholarship on information and cybernetic theory has quietly taken care to avoid. Shannon and Wiener themselves presented their work not as theories of information but as theories of communication. Shannon titled his work "A Mathematical Theory of Communication."[30] A few years later, at the infamous Macy Conferences on cybernetics, he would state, "My own model of information theory . . . was formed precisely to work with the problem of communication."[31] Wiener's title tells a similar tale: *Cybernetics, Or Control and Communication in the Animal and the Machine*.[32] In the book's final chapter, he argues that the key to the fundamental problem of social organization is "intercommunication."[33] In the subsequent book he wrote to popularize cybernetics, he reaffirms communication as the "cement" of society such that "society can only be understood through a study of the messages and the communication facilities which belong to it."[34] Throughout his work, Wiener was clear that information is to be construed as a function of communication, as a "content of what is exchanged," rather than as a function or operation in its own right.[35]

Information theory was born not as a theory of information, but as a theory of communication that simply presupposed information as the material it would transmit. In other words, so-called information theory is a theory of channels or carriers of information, but not a theory of information itself.[36] It was less an originary moment for the information society and more a contribution to what Orit Halpern calls "the relentless encouragement of future communications."[37] Writing one decade after the information theorists, philosopher Richard McKeon would note the scope of that relentlessness: "'Communication' . . . is a term which has spread in use and implications during the past two or three decades from the problems of mass media, public relations, and promotion, to include all practical and social problems."[38] What makes this claim interesting is the *all*. It had already come to seem as if the whole point of information was for it to be communicated, a point for which McKeon himself argued: "A truth which we do not succeed in expressing and communicating is ineffective."[39] Given such consequences of their work, Shannon and Wiener are better thought of as presupposing information for purposes of communication than as contributing to information's formation. In presupposing information itself, information theory was less the dawning origin of an information era and more a dusky consolidation of decades of preexisting information practices.

Information theory could do its crucial work at midcentury because it could presuppose reliable information technologies that were already there. Information theory could presuppose information because, for a few decades already, information had been a site of intensive technical, cultural, and otherwise practical labor. That work—the site of the emergence of information as a concern and problematization—is what we can attend to only if we shift emphasis from the midcentury theorists of communication who presupposed information to the early-century information technicians who had elaborated the vast mazes of data on which communications theory would try to impose order. By shifting attention to information before information theory, we thus illuminate some of the ways in which information was less the design of later theoretical consolidators and more a mobile mélange arranged by technicians who found ways to informationalize anything and everything, including people. Information theory is itself part of the flotsam of those practices, techniques, and subjectivities of information.

THE LIMITS OF COMMUNICATIVE DEMOCRACY FOR INFOPOLITICS

The recent history of political theory reveals a striking convergence in the later decades of the twentieth century around communicative theories of democracy. These theories emphasize the role of deliberation, conversation, and discursive interaction in democratic politics. Most leading versions of these theories operate within the paradigms laid out in the 1970s and 1980s by Jürgen Habermas's discourse ethics and John Rawls's conception of public reason.[40] In their broader outlines these recent paradigms were anticipated by at least one protocommunicativist theory elaborated in the 1920s (Peters's decade of communication's consolidation): that of pragmatist John Dewey in his famous debates with William James's former student Walter Lippmann.[41] I shall consider both of these theoretical moments in turn.

In a 2003 review essay summarizing deliberative democratic theory, Simone Chambers noted that "nearly everybody these days endorses deliberation in some form or other (it would be hard not to)."[42] The years following this then-true observation have seen numerous new directions in political theory positioned explicitly against deliberative democratic theory.[43] In light of these recent critical efforts, my argument may appear to be just one more such attempt to disrupt deliberativism. But unlike recent criticisms of deliberativism, my claim is not that we need to move beyond communicative conceptions of politics because of what they get wrong. My claim is that we need to be able to move beyond communicativism about politics (and only for some purposes) precisely because of what it gets right.

If deliberative democratic theory is a useful intervention in the recent history of normative political theory—and I believe that for the most part it is—then its utility is a function of the depths to which politics today presupposes previously formatted information as a stuff in need of communicative exchange. Deliberative democratic theory is thus not wrong to orient its political guidance around communication. But in requiring that communication always be the means of its own improvement, these theories unwittingly establish their plausibility on the basis of insulating communicative polities from the politics of information itself. Communicative polities, and the communicative conceptions of democracy that provide them with guidance, can only presuppose information.

My argument, in contrast, has been that to confront information as a political problem we must confront information itself, at the level of its formats, rather than taking information's formatting for granted. Information always formats first. Thus, any attempt to confront information as a pure function of communication is bound to leave intact those formats that any communicative process must presuppose.

One entrée to my argument is by way of an unexpected parallel between formalist conceptions of communicative democracy, especially Habermas's, and the above-discussed theories of communication developed by the so-called information theorists, especially Shannon.[44] Jonathan Sterne observes that the midcentury information theorists "were concerned with the process, rather than the content, of messages."[45] The same, of course, can be said of Habermas. On the basis of this similarity, I shall argue that Habermas's democratic theory remains unresolvable on the very conceptual core that constituted Shannon's problem—namely, their shared communicative proceduralism.

To spell out this parallel in more detail, consider again that Shannon sought a theory of communication channels competent to efficiently carry any and all information, irrespective of its specific content. Had he designed a communication channel that would selectively avoid certain kinds of content (a telephone system efficient with the adult human voice but slow to transmit machine tones or incapable of transmitting certain spoken words), his theory would have been at best a partial achievement. The theory would have yielded a mathematics of communication not yet rising to the challenge of information's own universalizability (as if only a human voice, but not an automated tone-emitting box, can carry information).

Shannon's technical proceduralism is replayed by deliberative proceduralism. The same formal strategy is central to both approaches (though the strategy also clearly operates on different levels in each, given that deliberative democracy is explicitly constrained to rational political communication and not generalized to communication as such). Their common strategy is this: both theories are designed so as to remain agnostic about the contents of communication by focusing only on procedures of communication. Shannon's communication channels are designed to be agnostic to informational contents or meanings. Habermas's communicative rationality is a design for a formal procedure that is agnostic about the substance of what gets communicated. Important, however, is that the very formal move whereby information engineers solved the technical problem of efficient information transmission becomes a site of structural inattention within the horizon of a political problematic. In both theoretical paradigms, purely proceduralist communication cannot address itself as such to the formatting of information but can only presuppose it. This is a feature where these presuppositions need not be addressed, as on a telephone network. But it is a bug where these presuppositions carry political consequences, as is the case for the design of information technologies of selfhood.

To better see how this bug plays out in deliberative democratic theory, consider Habermas's theory in more detail. His crucial contribution to deliberative democracy is the idea that "the central element of the democratic process resides in the procedure of deliberative politics."[46] Such deliberative procedures offer normative guidance, or confer validity, to the extent that, as Habermas states with his "discourse principle" of action, "all possibly affected persons could agree as participants in rational discourses."[47] This principle presents an idealized theory of communication such that it can yield genuine normative guidance for action. It is a crucial feature of Habermas's discursive ideal that it is oriented entirely by requirements of communication understood as a social exchange in actual fact. Habermas's communicative orientation declines any perspective enacting a solitary employment of a faculty of reason, rigorously avoiding what he once called "monological mock dialogue."[48] What, then, is involved in this orientation to actual communication?

The crux of Habermas's communicative orientation can be gleaned from his elucidations of the idea of "rational discourse" employed in the discourse principle quoted above: "'rational discourse' should include *any* attempt to reach an understanding over problematic validity claims insofar as this takes place under conditions of communication that enable the free processing of topics and contributions, information and reasons in the public space constituted by illocutionary obligations."[49] Habermas's discourse ethics—at its most general level at which it applies widely to all species of moral norms, including but not limited to democratic procedures—involves a communicative perspective that endorses any information whatsoever insofar as it is freely processed. This view raises crucial questions about how Habermas conceptualizes information such that it is capable of being freely processed.

On this point, however, Habermas's theory offers no guidance about the status of information presupposed by free communication. That is to say, Habermas ignores information as a site of politics. This is because he must so ignore it. Consider his commitment to the idea that any signal which is freely processed must count as information that can fulfill the starting conditions of rational discourse. This commitment results in a necessary inattention to the politics of information itself.

Concerning the information that is to be freely processed in ideal communicative discourse itself, we could always ask: How is this information formed? How is it formatted? What burdens and benefits are embedded in those formats? But these are questions that a political theory taking communication as the metric of its critique cannot possibly confront, because rules for communicative exchange must be rules that already presuppose the information that will be the contents of communication.

Is this, however, truly a problem? It may appear as if Habermas's communicative proceduralism could address itself to those formats that impede the communicative process. Formats play a significant role in the politics of communication, after all. In preceding all communication, they help structure communicative exchange. Thus it may seem as if this prior structuring is precisely what Habermas's communicative proceduralism is designed to address.

But this is only how things seem. Habermas's theory will address the prior structuring of formats only as a function of their downstream effect on communication.[50] In other words, it is only when formats distort communication that communicative proceduralism can countenance them. If formats initiate or reproduce other kinds of political or moral effects, a theory of communicative interaction must remain silent about them. Throughout this book and particularly in the preceding chapter, I have outlined how there is an independent politics of formats not parasitic upon whatever communicative interaction happens to transmit those formats.

The real problem, in fact, is deeper. Communicative proceduralism must also unwittingly reproduce any and all freely processed formats. Information that does not distort communication is precisely what deliberative democratic theory is designed to cultivate. Thus, if there is a politics of formats, and if that politics does not always show up as a function of rational communicative exchange, then we are in the midst of a political state of affairs that communicative democracies could only ever innocently reproduce but never critically interrogate. This conclusion points to a precise specification of what communicative theory cannot address but which any viable political theory of the contemporary needs to countenance: information perfectly capable of being freely communicated and yet nevertheless politically problematic or dangerous in other ways.[51] There is today an abundance, even an overload, of such information.

Consider as an example something as banal as your social media profile. Surely these profiles establish data that facilitate specific forms of communicative interaction. But they also do much else. They also function as locales for a politics of fastening that is irreducible to a politics of communication. They format for users how they should see themselves and one another (perhaps often innocently enough, but surely not always innocently, as when we consider how this formatting impacts the selfhood of preteens coming accustomed to its use). They also format users into data points that can be used to construct assessments for targeting them as subjects of consumer marketing or political campaigns.[52]

Consider another example proximate to the histories analyzed in preceding chapters. Personal information theft has been a sustained and serious matter of concern in informational societies since at least the late 1950s, even if sporadic panics can be traced back through the centuries. One of the most recent episodes of widespread concern was a credit reporting agency's exposure in 2017 of the Social Security numbers (SSNs) of more than half of all Americans. The consequences of this giveaway will be felt over the coming years as these assets flood black and gray markets for personal information.[53] The political problematic at play here—like other problems of advanced technological democracies—is not a matter of communication but of the information that precedes communication. Addressing the problem of personal information exposure cannot only be a matter of providing for valid discursive exchange, or the free processing of information. Rather, the very information that is subject to processing must be reformatted. Our paper-and-print SSNs have grown extremely brittle in the midst of digital duplication and dissemination. Some critics have thus proposed cryptographic SSNs.[54] One influence here may be the impressive rise of cryptographic currencies—if cryptography can gain traction in the context of property exchange, then perhaps it can also do work in the context of personal information exchange.[55] Whatever solutions get developed, my point is that any successful response to the problem of personal information theft will surely take place at the level of rational communicative interaction, and as well at the prior level of information design and formatting—a level where any theory oriented entirely around communication must remain silent.

The problems I am signaling are not simply a function of Habermas's proceduralism—they are an implication of the strict limits he imposes on his proceduralism by attending only to specifically communicative procedures. As such, a more promising path for a normative political theory of information to explore would involve a radicalized proceduralism that is focused on both processes of communication and the processes of information design that any and all communication presupposes. Such a theoretical paradigm is precisely what should be expected if I am right, in the argument developed above, that infopower is a distinctive modality of power layered on the biopolitical, disciplinary, and sovereign powers characteristic of a more familiar moment in modernity. For what such a diagnosis suggests is the need for a normative political theory that could also attend to the processes of informationalization that precede, and are invoked by, all communicative processes. Such a normative political theory, at its most promising, could abide the crucial procedural-ist stricture that we cannot decide in advance what communicative contents count in politics and, at the same time, could find a way to countenance the politics of data by attending to the ways in which information performs a political work of formatting. Such a normative perspective would not be opposed to Habermasian deliberative democracy, but it would necessarily extend beyond the limits of his essentially communicative orientation.

Perhaps, then, Habermas's communication-centric neglect of the politics of information is an innocent and reparable oversight. Even if so, it is probably also symptomatic of a long-standing distrust of information on the part of critical theorists, and even of political theorists more broadly.[56] In 1944, Habermas's predecessors Theodor Adorno and Max Horkheimer mocked the contemporary "flood of precise information [präziser Information] and brand-new amusements," suggesting thereby that data and entertainment are consumables on a par.[57] In 1936, Walter Benjamin lamented the rise of the newspaper, saying that "it is no longer intelligence coming from afar, but the information which supplies a handle for what is nearest that gets the readiest hearing."[58] In specifying just what it is that gets lost in information, Benjamin in 1936 sounded quite like Eliot in 1934: "Narrative achieves an amplitude that information lacks."[59] Where are the stories we have lost in the news?

A severe distrust of data may seem a strong starting point in the face of the politics of formats. But a refusal to countenance the gains of data is an invitation to irrelevance. A political theory of information must learn to occupy data technologies from within, rather than assuming that they can somehow (and how exactly?) be refuted. In this respect, Habermas's communicative rationality is certainly a decisive step forward from first-generation critical theory. But, I have argued, a politics of redesigning data from within requires far more than even communicative theory can comprehend. To move toward an alternative, I turn now to a different precedent for Habermas's communi-cativism: Dewey's 1927 The Public and Its Problems, a work Habermas cites favorably.[60] It is not in Dewey's communicativism, however, that we begin to find a political theory that confronts in full seriousness the power of data. It is, as it happens, in Dewey's prompt, the political analysis of Walter Lippmann.

Dewey's book is a visionary argument on behalf of participatory democratic deliberation: "The essential need . . . is the improvement of the methods and conditions of debate, discussion, and persuasion."[61] Such communicative matters are, Dewey urged, "*the* problem of the public."[62] Throughout the book, communication is positioned as the best means for the achievement of democratic community: "Communication can alone create a great community."[63]

If Habermasian democracy meets its information-theoretic accomplice in Shannon's technical formalization of communication, the broader vision of Deweyan democracy finds its mate in Wiener's synoptic cybernetics. Consider Dewey's view that the "bewildered" public is an effect of its disconnect from the administrative machinery of government.[64] This idea is strikingly similar to what Wiener would later describe as the very problem for which "feedback" is the solution.[65] In fact, Wiener's argument closely replicates three crucial features of Dewey's analysis. First is the idea that "small, closely knit communities" have actually been able to achieve adequate organization.[66] Second is the idea that organization, or "homeostasis," will be pressured in any "society too large for the direct contact of its members."[67] Third is the theorization of this pressure in terms of the "constriction of the means of communication."[68] For Dewey and Wiener alike, the fundamental problem of modern politics is that of improving our communicative media by adjusting them to what Dewey called the "invasion" of community life by "new and relatively impersonal and mechanical modes of combined human behavior."[69]

Dewey set a crucial precedent for subsequent democratic theory in championing "free and systematic communication."[70] Thus do contemporary political theorists turning to Deweyan democracy find themselves approximating the basic terms of Habermas's communicativism. Consider as just one example the echo of Habermas in Cheryl Misak's description of Deweyan democratic theory as envisioning polities in which "inquiry requires the unimpeded flow of information."[71] This certainly sounds attractive. But we ought not neglect that Dewey and Habermas also converge in their failure to confront the possibility that information itself can be a form of political impedance. Their signal idea of free communication cannot, by its very terms, countenance the politics inherent in any information that flows unimpeded. This results in a major problem (as argued above vis-à-vis Habermas): any vision of a communicative politics that requires unimpeded flows of information as the basic materials for deliberative participation can only entrench existing formats, sometimes for better, and sometimes for worse.

If that is right, then perhaps we should turn our attention away from Dewey's deliberative democracy and consider with more seriousness the political theorist whom Dewey took as his main interlocutor in *The Public and Its Problems.* Dewey's communicative communities were a response to Lippmann's arguments in his 1922 *Public Opinion* and 1925 *The Phantom Public.*[72] In the first of these books, Lippmann criticized one of Dewey's political heroes, Thomas Jefferson, as inapplicable to the age of the radio and the newspaper, arguing that "conditions must approximate those of the isolated rural township if the supply of information is to be left to casual experience."[73] This was precisely the conception of democracy that Dewey would seek to rescue five years later in writing of the great communicative community as an experience of "face-to-face intercourse."[74] Why Dewey thought that full communicative presence was even remotely plausible in the globalizing metropolises of the early twentieth century is difficult to say.

What is of particular interest in Lippmann's argument, however, is not its criticism of an outdated conception of localized communication but rather its prescient focus on information as the terrain of political governance. If Dewey anticipated later communicative conceptions of democracy, then Lippmann's *Public Opinion* anticipated an option that subsequent political theory, including his own follow-up in *The Phantom Public*, has left widely unexamined: the politics of information itself.

Public Opinion mounted an argument for such projects as "bring[ing] industry under social control [by] the machinery of records" and "devising standards of living and methods of audit by which the acts of public officials and industrial directors are measured."[75] It envisioned "a focus of information of the most extraordinary kind" supplying "measures" like those being inaugurated at agencies such as the Children's Bureau, which Lippmann admired for its work in bringing visibility to issues of infant health with a success "as if the babies had elected an alderman to air their grievances."[76] Lippmann's analysis of the work of information specifically brought to the fore the "technic" of the "experts" of modern democracies, who bring with them "each a jargon of his own, as well as filing cabinets, card catalogues, graphs, loose-leaf contraptions, and above all the perfectly sound ideal of an executive who sits before a flat-top desk, one sheet of typewritten paper before him, and decides on matters of policy presented in a form ready for his rejection or approval."[77] These data-focused techniques and methods can, argued Lippmann, furnish insights about such matters as "people who are not voters, functions of voters that are not evident, events that are out of sight, mute people, unborn people, relations between things and people."[78] Lippmann fashioned a fecund term for the pell-mell concerns of public data: they are "a constituency of intangibles" to which data can help extend political participation.[79]

Lippmann's technics of forms, measures, and audits was expressive of a politics of information that looked forward to, even if it did not arrive at, the problems shaping that world of data analytics in which we live today. He seemed, already in the 1920s, to have almost in sight the political pressure that could be exerted by an appraisal form, an identification card, or a pencil-and-paper psychological questionnaire. Lippmann even concluded *Public Opinion* with a call for a more robust education in the critical scrutiny of uses of information.[80] He thereby nearly brought into view the political and ethical valences of the designs and formats that would have been the functional technics of the machinery of record he sought as central to the future of democracy. Yet close as he came, Lippmann never quite gained the vision. His subsequent work was redirected toward a realpolitik worldview easily tainted by its associations with elitism. That later air of elitism would then be read back into Lippmann's earlier proposals for expertise in information technology.[81] This anachronistic misdirect helped obscure the promise of Lippmann's early anticipations of a politics of information.[82]

The politics of information remains today, one hundred years later, theoretically underexplored. In its stead we meet countless conceptions of the politics of communication, nearly all of them distrustful of information as such. In bringing into view the limits of these theories, I am not arguing against the importance of communication for politics, but rather questioning the idea that communication is the most essential aspect of political interaction such that interrogating communication would be sufficient for critical political theory. Here we could finally take a clear cue from the obscure messaging of Deleuze and Guattari: "We do not lack communication. On the contrary, we have too much of it. We lack creation. We lack resistance to the present."[83]

The concerns about the prioritization of communication I have been raising seem to share a common target with recent tendencies in political theory to cast doubt on the aspirations of communicativism. However, much of this recent criticism looks toward alternatives that are distant from what I have been arguing for. The most prominent current alternative to communicativism is perhaps that of a politics of aesthetics and affects. I could not dismiss the aesthetic. Like you, I have felt its blisses: the telltale tingle between the shoulder blades, the grandeur of the gray peaks catching first golden dawn, all those fortnightly romps when we lost ourselves in tiny, ecstatic basements where sweaty, high bodies trembled to music so impossibly vibrant that only we could have embraced it as purely as we did. Yet we secretly knew even in those moments that what was then so very heaven could not be a viable politics and a buildable world. Reconstructing politics requires not just the quick lures of the aesthetic but also lasting designs we can trust.

The alternative to communicativism that I find compelling—and this book has attempted to convey why—is that of a politicized technics.[84] The technics of information may lack the austere rigor of communicative rationality and the blooming perfections of the passions. But austerity and perfection are both liabilities in politics. By contrast, a critical interrogation of the complexities and contingencies of technics can bring us into decisive confrontation with the operations of power, and therefore with the possibilities of resistance.

The Future in Data: Toward Resistant Informatics

What is resistance in a present saturated by data? What would it mean to mount resistance to a politics of information? If this book has demonstrated anything, it is that it would be a nonstarter today to construe political resistance in terms of resistance to information as such. What could it even mean to be against information today? To take a loud stand against data, or even just quietly reject it? The very idea is incoherent, impossible, incredible. We live within a data episteme and under a power of information. We are informational persons.

If information has become our functional universal, at least for the foreseeable future, it need not therefore function as our unquestioned presupposition. Resistance can be conducted *within* the operations of infopower: a resistance to *this* kind of fastening, a resistance to *that* kind of canalizing and accelerating, and a resistance that mounts these actions by repurposing and releveraging information for alternative designs. This would be a resistance of occupation, contestation, and transformation. It would be a politics of critique for which there are numerous salient historical precedents—for example, William James's late nineteenth-century efforts in self-transformation in the midst of a flood of probabilities defining the self.[85] Yet as much as James's problematic of probabilities anticipates later forms of formatting, his flood was not exactly what we are saturated in by our contemporary condition of information.

This is why it is so crucial for us to seek the specific historical precipitants for our present through which we can locate contemporary sites of infopolitical occupation, contestation, and transformation. This is why this book has sought to excavate dynamics of information just prior to their consolidation in formats that are now seemingly incontestable. And lastly, this is why we must conceptualize the politics of information in a manner attentive to its specific functions, designs, and operations.

It cannot be the charge of political theory to state in advance what shape such an alternative informatics of resistance must take. That would be to assume the role of the prophets, programmers, and paranoiacs who indulge our fantasy for algorithmic certainty. Declining that role, political theory can glimpse resistance practices as they begin to emerge, and then perhaps inquire into the conditions by which they might manifest in greater intensity. In the present moment, I believe, we find ourselves only at the front edge of gaining the occasional glimpse of such emergent counterconducts.

When we can see it, what we glimpse is the relentless yet frustrated work of a whole assembly of resistant informatics: the work and care and love of a multitude of auditors, sharers, leakers, hackers, sousveillers, obfuscators, and anonymizers.[86] The theorists attentive to this diverse set of resistances are all concerned with a problematization of infopolitics. And if these theorists are right that there is a politics put into operation by the informational accoutrements with which we ceaselessly interact, then it follows that we would do well to situate emergent styles of resistant informatics in sites that seem more mundane than what we think of when we consider the work of politics. Such humble sites may very well be the primary terrain upon which will be fought out the struggles over informatics that are sure to characterize the remainder of the twenty-first century. The sites I have in mind are those where we find the formatting of information systems, information technologies, and of course information itself.

It may be a design lab. It may be a code studio. It may be a tech incubator. It may be an engineering firm. It may be a seminar table around which regularly gather a small team of collaborators, all of whom know that there is already much at stake in these matters. Or it may be, as high-tech mythology loves to pretend, somebody's mom's garage. These are the sites where formatters, designers, and developers build the information systems that form the basements beneath our lives. What decisions are today's formatters making? Which of their techniques are increasingly formatting us today? Who among them considers the downstream effects of their designs? Who interrogates those effects with the meticulousness demanded by the politics of formats being enacted across these sites?

Today's information designers, just like the information architects of yesteryear, may be doing more to format our futures than they (and we) might think. There are countless designers who at this very moment are setting into motion tableaux, protocols, databases, and other informatics techniques that may come to have lasting influence for generations. In their formats (which are also our formats) are already contained decisions and pathways that will entrench specific informational subjectivities for decades to come. This could happen for tomorrow in just the way that you and I are today entrenched in formats that were designed in our shared past. Today's obligatory information technologies are the downstream detritus of data apparatus initialized a century ago. Thus we are obliged to ask ourselves about the politics of to-day's designs as they are becoming tomorrow's formats. What expectations and requirements will be ushered in by the informational apparatus we are now assembling? What will be the formats through which everyone conducts themselves in the next decade, and in the next century? Are we even in a position to recognize the entrenchments into which we are swiftly canalizing our futures? Or are we more likely to be formatted in the same way again—that is, relatively unthinkingly?

In considering these questions, we need not pretend that we can arrive at anything like a clear moral conclusion. Technology is not morally transparent so much as it is politically fractured. Informatics techniques have been put to patently unjust uses at the very same time as being employed for causes plausibly claiming to pursue a more perfect justice. The attention I have hoped to bring to yesterday's information technicians as an object lesson for tomorrow's formatters (and formattees) yields not a comforting morality play, but rather a complicating genealogy of sociotechnics in their ongoing and fragile development.

The genealogy I have offered here excavates a fragile yet durable past. The genealogist is an inquirer who discerns fissures on the surface of the crackling furnace of the past in which our narrow now was forged. By inserting ourselves into those fissures, we can perhaps redirect enough of the hot heat so that tomorrow may be transformed. Where do we resist? In the forms, formats, and information. How do we assemble the competencies to resist these forms and formats? By learning how to reformat them. By understanding how to redesign them. By interrogating the manifold technologies with which they have been designed and redesigned. Thus does resistance for the sake of the forthcoming future require critical engagement with the proximate past.

This book was written out of the hope that such a complicated relationship between our politics and our past, as old an idea as that is, nevertheless still matters to the present. "So we beat on, boats against the current, borne back ceaselessly into the past."

Acknowledgments

Books are always of complicated genesis. The ideas for this one have been percolating for at least a decade. They did not begin imprinting onto paper until about half as many years ago. If their impress has come to resemble an argument, it is owing to many individuals and, even more important, to a small few groups of individuals.

The field of focus here first gained plausibility as a project in the midst of a chance opportunity to spend time with an inspiring group of graduate students and postdocs (of which I was one) congregating around Paul Rabinow's Anthropology of the Contemporary Research Collaboratory (http:// anthropos-lab.net/index.html) in Kroeber Hall at University of California, Berkeley. The concepts I was experimenting with back then are hardly recognizable in this book, but the book is nonetheless the end point, for the nonce, of what I began to take seriously in that context.

In part modeled after that collaboration, a second context of genesis has been the University of Oregon Critical Genealogies Collaboratory (http:// uocgc.blogspot.com/), which has involved a number of supportive colleagues and inspiring graduate students. Especially deserving of thanks are the fabulous group of graduate students who joined me during 2016–17 for collaborative research on the genealogy of birth certificates: thank you, Bonnie Sheehey, Patrick Jones, Laura Smithers, Sarah Hamid, and Claire Pickard.

A third group that has proved equally crucial to sharpening, but also broadening, the thinking in this book is the Critical Genealogies Workshop (http:// criticalgenealogies.weebly.com/), an approximately annual event I have had the good fortune of organizing alongside Verena Erlenbusch-Anderson of the University of Memphis. My thanks to you all: Verena, of course, and as well

Natalie Cisneros, Ray Dahl, Andrew Dilts, Simon Ganahl, Robert Gehl, Jesse Houf, Stephanie Jenkins, Sean Lawson, Mary Beth Mader, Ladelle McWhorter, Thomas Nail, Kevin Olson, Brad Stone, and Perry Zurn.

Beyond the regular conversations afforded by these groups, numerous occasions for public presentation of this work helped me further sculpt the ideas here. These occasions and the individuals who made them possible, along with others who extended generous occasions for thinking there, were the University of Washington Simpson Center for the Humanities (Gaymon Bennett, Meg Stalcup, and Alison Wylie), the Frontiers of New Media conference at the University of Utah (Robert Gehl and Sean Lawson), the Penn State University Rock Ethics Institute (Nicolae Morar), the University of Memphis Department of Philosophy (Verena Erlenbusch-Anderson), the University of Portland Department of Philosophy (Jeff Gauthier), the "Was heißt: Foucault historisieren?" event at Universität Zürich (Simon Ganahl), the University of Nevada–Reno Department of Philosophy (David Rondel), the UCLA Experimental Critical Theory Program (Davide Panagia), the University of Utah Department of Communication (Robert Gehl, again), the Oregon Humanities Center at the University of Oregon, the Social and Personality Psychology Research Group at the University of Oregon, the Australasian Society for Continental Philosophy (Paul Patton, Sean Bowden, and, for always-insightful questions, Amy Allen), the Deakin University Department of Philosophy (Sean Bowden), the Kingston University Centre for Research in Modern European Philosophy (Catherine Malabou), the Oregon State University Department of Philosophy (Stephanie Jenkins) and Center for the Humanities (Joshua Reeves and Christopher Nichols), the Columbia University Center for Contemporary Critical Thought (Bernard Harcourt), the Evergreen State College (with much gratitude to Eirik Steinhoff for the opportunity of a wonderful return to my alma mater), Emory University's Oxford College (Erin Tarver and Josh Mousie), and sessions at the American Political Science Association, the Western Political Science Association, the American Philosophical Association's Pacific Division, and the Critical Genealogies Workshop (as noted above). This work was also improved through publication of earlier drafts of some material. Portions of chapter 4, here revised, that appeared as "Infopolitics, Biopolitics, Anatomopolitics: Toward a Genealogy of the Power of Data," *Graduate Faculty Philosophy Journal* 39, no. 1 (2018): 103–28. Ideas from the introduction and chapter 5, also revised, are due out as "Information before Information Theory: The Politics of Data beyond the Perspective of Communication," *New Media and Society* (forthcoming, 2019).

Beyond all these opportunities for shared engagement, numerous individuals read this material and offered feedback. My deepest debts are to Verena Erlenbusch-Anderson for reading every chapter, some of them in multiple drafts. Thank you also for the inspiration of your work.

A small group gathered at a manuscript workshop in April 2017 to offer feedback on what was then the first full draft of this project: my thanks yet again to Verena, and to Daniel Rosenberg, Gerald Berk, Nicolae Morar, and Chris Penfield. Later that spring, students in my graduate seminar "Data Genealogy" read most of a draft of the book and offered valuable feedback. Numerous other University of Oregon colleagues helped me think through matters of this book as it developed, including, most notably (and yet certainly not only), Michael Allan, Kate Mondloch, Nicolae Morar, Daniel Rosenberg, and Rocío Zambrana.

Some months later, with a push of encouragement from Arnold Davidson for which I am grateful, I sent off a book proposal and received solid advice on the penultimate manuscript from Davide Panagia and an anonymous reviewer for the University of Chicago Press.

Along the way, various individual chapters or chapter pairs benefited from feedback by Wendy Hui Kyong Chun, Robert Gehl, Christopher Kloth, and Natasha Dow Schüll. Right at the end, Caroline Koopford read the whole thing and offered customarily perfect advice.

Editorial attention in the later stages of finalizing the book helped everything run smoothly. My thanks to Elizabeth Branch Dyson for immediate support of this project, to Dylan Montanari and Tamara Ghattas for keeping things on track, to Johanna Rosenbohm for a keen eye on the copy, and to the rest of the professional staff at the University of Chicago Press working on this project. Additional thanks for late editorial assistance are due to the University of Oregon Department of Philosophy, and in particular Kit Connor and Rebecca Saxon, who provided needed proofing as the project neared completion.

I am also grateful to numerous libraries, librarians, archives, and archivists. Chief among them are the fine staff of the University of Oregon Libraries, especially Mark Watson, who helped me navigate my much-loved "fourth-floor stacks" when they were unexpectedly, even if only temporarily, moved off-site some sixty miles away. I thank the University of Reading Neurath archive for their materials, including the "Genealogy of Isotype" image here reproduced under license from the center. I thank the University of Akron Cummings Center and Lizette Barton for a treasure of resources from the history of psychological testing, and in particular for providing access to the original "Personal Data Sheet," here reproduced under license from the Psychology Tests

Collection, Archives of the History of American Psychology, Drs. Nicholas and Dorothy Cummings Center for the History of Psychology, University of Akron.

A number of units at the University of Oregon sponsored this project with fellowships or project grants of various kinds that afforded me much-needed research time and in some cases teaching projects directly related to the book: the Wayne Morse Center for Law and Politics Resident Scholar program, the Oregon Humanities Center Faculty Research Fellowship program and Wulf Teaching Fellowship award, a Williams Fund Instructional Proposal grant, a Research Innovation and Graduate Education Summer Stipend Award, the Provost's Office, the College of Arts and Sciences Humanities Summer Stipend program, the Office of the Vice President for Research and Innovation Faculty Research Award program, and, from the same office, an Incubating Interdisciplinary Initiatives program sponsoring an inspiring research collaboration with Dr. Jun Li of our Computer and Information Science Department.

I am deeply grateful to all the above-named persons and organizations, as well as for the opportunity to register that gratitude here in what will become searchable and search-engine-archived text, thereby making my appreciation a little more strangely real, at least according to the terms of our world as it is for now.

My deepest gratitude over these years of writing goes to my wife and our children. And perhaps I shall leave the details of this most important little bit unsearchable, affording the machines and minds that do not already know your names the opportunity to later database them.

Agate St. and 19th St.
Eugene, Oregon
January 2019

Figures

Notes

Citations indicate original publication dates for reprints of an edition printed in English, as in "Wiener (1948) 1961, 3"; for chapters in books that were first published elsewhere, as in "Haraway 1991 (orig. pub. 1985), 161"; and for foreign-language works cited in English translation, as in "Foucault 1995 (orig. pub. 1975), 28."

Introduction

1. Neurath 2010, 5, 126.

2. Neurath 2010, 113–14.

3. Neurath 2010, 81.

4. On the unity of science, see N. Cartwright, Cat, Fleck, and Uebel 1996, part 3, 167–252.

5. The excellent book by N. Cartwright, Cat, Fleck, and Uebel expresses a typical bias in the history of philosophy in that it notes Neurath's work on Isotype and information design only in passing (1996, 63–72, 84–85, 92, 182).

6. Neurath 1936, 111.

7. Neurath (1937) 1973, 224.

8. Turovsky 2016. For a media genealogy of Google Translate as evidencing a universalism of "uniform multilingualism," see Ramati and Pinchevski 2017.

9. The quoted language is Google's mission statement at the time of writing; Google, "Our Company," accessed November 22, 2017, https://www.google.com/intl/en/about/our-company/.

10. Vismann 2008 (orig. pub. 2000), 112.

11. See Foucault 1990 (orig. pub. 1976), 18–21, 58–69.

12. Hacking (1990) 2006, 34.

13. See Foucault 1990 (orig. pub. 1976), 25, 144, 146.

14. Schüll 2016, 328, 326, 325.

15. Harcourt 2015, 1.

16. Gitelman 1999, 11.

17. For another such genealogical approach, taking personhood in general as its focus, see an in-progress project summarized in McWhorter 2017.

18. On habitual new media, see Chun 2016.

19. Foucault 2011 (orig. pub. 1983), 5.

20. Schüll 2018, 28, 35. My language of constitution and mediation here borrows a formulation offered by Schüll at a presentation at the Columbia Center for Contemporary Thought, Columbia University, October 27, 2017.

21. I refer here to Donna Haraway's (1991 [orig. pub. 1985]) conception of cyborgs and the seemingly distant but actually resonant idea of the extended mind developed by Andy Clark and David Chalmers (1998).

22. Why so casually dispense with the metaphysics of the matter? In short, because I find that metaphysics as a method too easily produces what Sherwood Anderson, writing in 1919, described as "grotesques." Discussing what too often happens to the "hundreds and hundreds" of truths composing our lives, Anderson observed that "the moment one of the people took one of the truths to himself, called it his truth, and tried to live his life by it he became a grotesque" (S. Anderson [1919] 1995, 6). I appeal only to Anderson's poetic portraiture because I have no brief to take a stand against metaphysics. I simply wish to pose philosophical questions beyond the limits of metaphysics. To do this I am content to leave it to the metaphysicians to debate what is and is not essential, that is, what is grotesque.

23. One example is the case of "the Erased," a group of more than twenty-five thousand former Yugoslav citizens residing in Slovenia whose identity records were erased when Slovenia declared independence in 1991; see Hervey 2017. Another example is the case of Alecia Faith Pennington, who was denied by her parents a birth certificate, and subsequently all other identity documents, until the 19-year-old's story finally persuaded a Texas state representative to forward a bill enabling delayed birth certification in cases like hers; see Radiolab 2016, with thanks to Natasha Dow Schüll for pointing me to this podcast.

24. With this qualifier, I dissent from the prominent general definition of information (GDI) defended in Floridi 2003. This definition is too restrictive in its requirement that information be semantically "meaningful" (2003, 42)—for this rules out what gets communicated in such contexts as internet packet transmission.

25. If I had to offer a definition, I would define information as any mass of discrete units in relations of difference such that the differences are capable of storage, analysis, and retrieval in sites beyond their context of origination. This is obviously a very permissive definition that gives information an enormously wide scope. But this permissiveness is a merit insofar as we are extraordinarily indulgent about what we are willing to apply the concept of information to. Indeed, almost anything can be information today; any viable definition of information must respect that. This helps explain why, throughout the book, I rely on the colloquial manner of treating data and information as everyday synonyms. A distinction may be upheld in some technical contexts according to which data is raw and information is always a cooked interpretation. But that distinction's slipperiness can be recognized twice over: first, in the fact that its defenders find themselves having to argue for it at all (just do a simple online search for both terms), and second, in the immense theoretical challenge facing the idea that there can be any such thing as raw data at all.

26. Foucault 1978, 226–27.

27. Foucault 2008 (orig. pub. 1979), 3.

28. See my previous discussion, Koopman 2013, 228–41.

29. My approach here resonates with Kittler's project for an analysis of "the network of technologies and institutions that allow a given culture to select, store, and process relevant data" (1990 [orig. pub. 1985], 369). More generally, my reliance on a technical model as an organizing device borrows from Hu's use of the OSI model of protocol layering for networking technologies (2015, xxv).

30. On algorithms, see Gillespie 2014 and my further discussion in chapter 2.

31. Amoore 2013, 51.

32. Amoore 2013, 51.

33. My uptake of Foucault in these terms is further explained below in the section of method, where I emphasize the difference between Foucault's analytics (e.g., genealogical method) and his concepts (e.g., biopower or discipline); see p. 23.

34. Foucault 1982, 241.

35. On Foucault's decomposition of power, see Koopman 2017b.

36. I use the term *assembly* throughout the book as a technical term roughly proximate to Foucault's category of the *dispositif*. Foucault's most generative description of his term is found in a 1977 interview: "What I'm trying to pick out with this term is, firstly, a thoroughly heterogeneous ensemble consisting of discourses, institutions, architectural forms, regulatory decisions, laws, administrative measures, scientific statements, philosophical, moral and philanthropic propositions—in short, the said as much as the unsaid. Such are the elements of the *dispositif*. The *dispositif* itself is the network [*réseau*] of relations that can be established between these elements. Secondly, what I am trying to identify in this *dispositif* is precisely the nature of the link [*lien*] that can exist between these heterogeneous elements. . . . Thirdly, by *dispositif* I understand a sort of—shall we say—formation which has as its major function at a given historical moment that of responding to an urgency [*urgence*]. The *dispositif* thus has a dominant strategic function" (1980 [orig. pub. 1977], 194–95; translation modified where indicated). I use "assembly" not as a translation of Foucault's *dispositif* (which has been translated as "deployment" and also quite misleadingly as "apparatus," and which bears comparison to Deleuze's notion of *agencement*, often misleadingly translated as "assemblage"), nor as a proxy, but as a term of resonance. For further on the *dispositif*, see Panagia 2019.

37. Haraway 1991 (orig. pub. 1985), 161.

38. Haraway 1991 (orig. pub. 1985), 164.

39. Kitchin 2014, 192; cf. 129, 151, 158, 163.

40. See, for instance, Haraway 1991 (orig. pub. 1985); Kittler 1999 (orig. pub. 1986), 2006 (orig. pub. 1996); Peters 1988, 1999; Bowker 1993; Galison 1994; Hayles 1999; Terranova 2004b; Clarke 2010; Halpern 2014; Kline 2015; Hu 2015; and in a different vein but with a focus on the same periodization, see Erickson et al. 2013.

41. Wiener (1948) 1961; Shannon 1949 (orig. pub. 1948).

42. There were two distinct branches in American information theory in the 1940s and 1950s: the cybernetics branch developed by Wiener and the engineering branch forged by Shannon. These branches converged on many points, but there is an important difference in that the cybernetics branch was much more ambitious in light of the efforts of Wiener and others to connect their concept of information to biological and psychological concepts not featured in Shannon's branch. My focus here is limited to information theory, and thus to only the information-theoretic aspects of cybernetics as pursued by Wiener and others; for broader accounts of cybernetics, see Heims, 1993; Pickering 2010; and Kline 2015.

43. See Orwell 1949.

44. Halpern 2014, 48.

45. Baldwin (1972) 2007, 52.

46. Deleuze and Guattari 1991, 49; they offer this as "the first principle of philosophy" (7).

47. For a detailed presentation of the difference between Foucault's *analytics* (or *methods*) and his *concepts*, see my essay trio Koopman and Matza 2013; Koopman 2014; Koopman 2015b.

48. On "the contemporary" as an analytical category, my approach is indebted to the work of Paul Rabinow; see Rabinow 2008, 2011; Rabinow and Stavrianakis 2014.

49. Du Bois (1903) 1989, 174.

50. See Koopman 2013.

51. On my conception of pragmatism, see Koopman 2009; on critical theory, see Koopman 2013, chap. 7.

52. Foucault 1982, 326. See discussion of this term by Lorenzini (2016); see also more recent research in this vein, on "kinds of people" by Hacking (1986) and Davidson (1990); and on "figures" by McWhorter (2009) and Dilts (2014).

53. Terranova 2004a, 51; cf. 2004b, 37.

54. Terranova 2004a, 52.

55. Foucault 1995 (orig. pub. 1975), 28.

56. Foucault 1973 (orig. pub. 1963), xvi, xvii.

57. Hacking 2002 (orig. pub. 1979), 92–93; see also Dreyfus and Rabinow 1983; Deleuze 1988 (orig. pub. 1986), 15.

58. See Farge and Foucault 2016 (orig. pub. 1982).

59. Farge 2013 (orig. pub. 1989), 28, 30.

60. See Kittler 1999 (orig. pub. 1986), 2010 (orig. pub. 2002). Further work in the vein of media archaeology includes Vismann 2008 (orig. pub. 2000); Zielinski 2002; Krajewski 2011 (orig. pub. 2002); Ernst 2013; Huhtamo 2013; and in a related vein Kafka 2012 and Gitelman 2014. With respect to this vein of scholarship, my own approach is probably best described as a media genealogy, consonant with a recent introductory article by Monea and Packer (2016).

61. Kittler 1982, 10.

62. Kittler 1999 (orig. pub. 1986), 230.

63. Bassett 2015, 192. See also Wellbery on Kittler's "post-hermeneutic criticism" (1990, vii); Krämer on his "media beyond the register of signs" (2006, 93); Winthrop-Young on his "analytical shift from the inside to the outside, from human truths and messages to rules, conditions, and (technical) standards" (2011, 53); and Hansen on his "suspension of the evaluative category called 'meaning'" (2015, 221).

64. See, for example, Bouk 2015; and Yates 1989, 2005.

65. See Koopman, forthcoming (building on Kay 2000; Chun 2011b, chap. 3; Leonelli 2016). I argue there that the central stakes of genetic conceptions of selfhood are a function not of the supposed reductionism of genetic determination but of the plurality of ways of operationalizing genetic-behavioral correlations. For contemporary data science views, see Cartwright, Giannerini, and González 2016.

66. My method here thus borrows from Foucault's (1971) distinction between genealogies of "emergence" and histories of "origins."

67. Headrick 2002.

68. Noiriel 2001, 31 (see 28 for the law).

69. Noiriel 2001, 31–32.

70. Scott, Tehranian, and Mathias 2002, 33.

71. Hacking (1990) 2006, 19.

72. On both Fichte and Bentham in this context, see Caplan 2001, 49, 65.

73. Bentham 1838, part 3, chap. 12, problem 9, 557. Bentham's *Principles of Penal Law* was probably first composed in the later 1770s. Its first publication was not until the French edition of 1802 in Dumont's collected three-volume *Traités de Législation* by Bentham (where it is split

across volumes 2 and 3). Its first English publication was Bowring's collected 1838 *Works of Jeremy Bentham* (where it appears in volume 1).

74. Bentham 1838, 557.

75. Bentham 1838, 557.

76. Bentham 1838, 557.

77. Bentham 1838, 557. See discussion of Bentham's proposal in Caplan 2001, 65.

78. Bentham 1838, 557.

79. Bentham 1838, 557. Bentham recognized that such a proposal would meet with objections and in response suggested a public opinion campaign led by "imprint[ing] the titles of the nobility upon their foreheads" (557); it is hard to not believe that Bentham is here joking.

80. Fichte 2000 (orig. pub. 1796–97), 291/254.

81. Fichte 2000 (orig. pub. 1796–97), 295/257.

82. Fichte 2000 (orig. pub. 1796–97), 295/257.

83. For instance, a recent study by Valentin Groebner (2007) would seem to dispute my historiography in its claim that our contemporary identity papers "are in fact thoroughly medieval" (8). I do not dispute Groebner's historical accuracy; I only point out that my thematic focus here is simply quite different from his. He is interested in "a history of objects" whereby he can trace current identity papers back to their earliest precursors and prototypes (12). I am less interested in materialities and objectivities, and much moreso in practices. The difference is crucial. Groebner traces our identity papers back to the 1400s when soldiers (172), religious pilgrims (175), beggars (178), and other classes of travelers were obliged to carry various kinds of paperwork that Groebner regards as a precursor of our modern passports. But, as Groebner himself notes (9), these early "passports" preceded modern nationality, which is the authority for passports as we use them today. Thus it remains unclear exactly what kind of precursor these earlier papers are—a point noted by Craig Robertson in his observation that there are multiple "discontinuities" between early modern "passports" and our contemporary passports (2010, 15). For my purposes, there is at least one crucial disanalogy from my study in chapter 1 of birth certificates and social insurance registration numbers: Groebner's premodern passports were required only of specific classes of persons. Today, nearly every traveler must carry a passport to legally cross certain borders. Crucially, however, many citizens still never travel internationally (the US State Department estimates that less than half of Americans have a valid passport). For all the attention the passport receives, it is not designed as a universalizable information technology of the sort that interests me. By contrast, birth certificates and registration numbers are today functional only if they can be presumed to be universal. That is why we all have (or very much want to have) them. It is that all-pervasiveness of information that distinguishes my project from Groebner's, and makes interesting histories like his something other than an objection to what follows here.

84. These earlier sciences were of course failures, and that is the crucial difference my periodization is meant to help bring into focus. On how such earlier projects failed, see discussion by Carson (2007, 102) and Gould ([1981] 1996).

85. See Engerman 2012.

86. Browne's project excavates from the eighteenth-century management of racialized enslavement a range of cases exhibiting the operations of a disciplinary surveillance she connects to Foucault's work (2015, 12). For example, one of Browne's chapters is devoted to the *Book of Negroes*, which was "the first government-issued document for state-regulated migration between the United States and Canada that explicitly linked corporeal markers to the right to travel" and

"the first large-scale public record of black presence in North America" (25, 70). The book's entries detailed passengers' physical descriptions, ages, and conditions of former enslavement in a way that suggests its use for identification in cases where a passenger's status (as free or enslaved) was (or could later be) under dispute (84–86). Browne's argument is that the *Book of Negroes* shows us how "the body made legible . . . has a history in the technologies of tracking blackness" in a way that forms "an important, but often absented, part of the genealogy" of what she calls "racializing surveillance" (70, 12). Technologies of racialization contrastively figure in my argument not by taking the reader back to early antecedents in the service of slavery but rather to early twentieth-century moments of racialization in the service of other (often more subtle) forms of hierarchy. These later moments markedly contrast from earlier iterations in the scale of their racializing informatics. The *Book of Negroes* functioned much as we would expect of an early precursor: it recorded only three thousand names and was in use for only seven months (83, 86). The input forms, queryable databases, appraisal algorithms, and visualizing reports (i.e., redlining maps) that I study in chapter 3 collected residential and financial data on millions of persons and remained in use for decades in such a way that they continue to underpin contemporary mortgage markets.

87. For early twentieth-century historiography of information, see Yates 2005, pt. 1, and Bouk 2015 on life insurance informatics; Purdon 2016 on literary representations of informatics; Mindell 2002 on communications and ordnance technologies in the interwar period; Lemov 2005 on interwar social sciences; Igo 2007 on statisticalizing social survey; Robertson 2010 on the American passport as an identity document in the 1930s; and, much broader in scope, see the ambitious early synthesis by Beniger 1989.

88. Purdon 2016, 4.

89. Bouk 2015, xviii, 63; Purdon 2016, 57.

90. On the English registries, see Szreter 2012.

91. See Cole 2001 and Sengoopta 2003.

92. Hu 2015, xxi. Hu dates this American universalism as a development from the "1950s on," whereas I would regard it as already consolidated by the period under survey here.

93. On the success of the American internet project in competition with other national projects, see Abbate 1999, 177–80, 208–12; and Mowery and Simcoe 2002.

Chapter 1

1. Cook 2017.

2. California Legislative Information. Health and Safety Codes, 2001, "Health and Safety Codes," div. 102, part 1 (Vital Records), chap. 3 (Live Birth Registration), section 102405, accessed August 15, 2017, http://leginfo.legislature.ca.gov/faces/codes_displaySection.xhtml?sectionNum =102405.&lawCode=HSC.

3. Los Angeles County Department of Public Health, "Requirements for Registering Out-of-Hospital Births Occurring in Los Angeles County Vital Records Jurisdiction," accessed August 15, 2017, http://www.ph.lacounty.gov/dca/vrecords/Eng_HomeBirths.pdf.

4. Caplan 2001, 50. Our current two-name system had stabilized across most of Europe by the twelfth century (Wilson 1998, 150), but the technique of a fixed and unique name was by no means widespread. For one, names tended to not be unique—in mid-seventeenth century London, just ten first names accounted for almost three-quarters of all names (Wilson 1998, 187). Second, everyday usage did not always conform to names as entered on church and

other registers (Wilson 1998, 234), and many persons even assumed different names for different contexts—one name among family, another among in-laws, and yet another with members of their age group (Wilson 1998, 235). Third, spelling and orthography were not standardized (Wilson 1998, 241).

5. Caplan 2001, 63.

6. Cited in Wilson 1998, 243.

7. Cited in Wilson 1998, 280.

8. Cited in Wilson 1998, 245.

9. Cited in Robertson 2010, 43.

10. Scott 1998, 65; see also Scott 1998, 64–71; Scott, Tehranian, and Mathias 2002.

11. Scott's preferred term is "manipulation"; see Scott 1998, 183.

12. Scott 1998, 2.

13. Scott 1998, 2.

14. Scott, Tehranian, and Mathias 2002, 14, 37.

15. See Scott, Tehranian, and Mathias 2002, 22. A more detailed history is offered by Littlefield and Underhill, who also describe the project's failure (1971, 41). Surprisingly, Scott, Tehranian, and Mathias (2002) do not note the poor results of the project despite the case fitting Scott's (1998) general theme of failure.

16. For an excellent analysis of resistances in these contexts, but of course much later, see Currah and Moore 2009 and 2015 on the role of birth certificates in gender identity and trans politics.

17. See the collections by Caplan and Torpey (2001) and Breckenridge and Szreter (2012); see also Higgs 2011, 2004; Robertson 2010; Lyon 2009; Amoore 2008; Agar 2003, 2001; and Torpey 2000.

18. On Scott, see Caplan and Torpey 2001, 1; and Szreter and Breckenridge 2012, 8.

19. Szreter and Breckenridge 2012, 1.

20. Fahmy 2012, 335.

21. US Children's Bureau 1919, 3.

22. US Children's Bureau 1919, 3.

23. US Children's Bureau 1919, 3.

24. For these phrases, see American Medical Association 1915, 154; American Child Health Association 1927, 6, 17. In 1929 the latter organization referred to "accurate records of births" as "the first step in the 'Bookkeeping of Life'" (American Child Health Association 1929, 17). For the "bookkeeping of life" phrasing, see also A. T. Davis 1926, 26. The first use of the general term that I find is of the "bookkeeping of humanity" formulation by J. N. Hurty, secretary of the Indiana Board of Health, who used it as a metaphor for any vital statistics in a 1910 address (Hurty 1910a; 1910b). This formulation predates "human bookkeeping" (note, though, that Google Books lists several incorrectly dated positives).

25. Pearson 2015, 1146.

26. Pearson 2015, 1163; for other recent historical surveys of birth certificates, see Marshall 2012; and Landrum 2015.

27. Census pamphlet nos. 100, 101, 103, and 104, all published in 1903.

28. US Census Bureau 1903a. Though focused on death registration, the argumentative structure adumbrated in this pamphlet would later be directly applied to birth registration, as in pamphlet no. 104, *Registration of Births and Deaths* (1903c).

29. US Census Bureau 1903a, 10–14.

30. US Census Bureau 1903a, 11.

31. US Census Bureau 1908, 7.

32. Scott 1998, 2 (and see discussion above).

33. See, for example, the work of the American Association for the Study and Prevention of Infant Mortality, which later folded into the American Child Health Association, who emphasized the legal argument (Ehler 1912, 102); the emphasis on the public health argument by the American Medical Association (1912, 1); and both arguments in a Children's Bureau publication (1913).

34. Commonwealth of Pennsylvania 1852, 2 (passed at the end of the 1851 session, the act was included in the 1852 report).

35. Balfe 1918, 776. On the history of Connecticut's registration system, see Balfe 1918; on Massachusetts's, see Gutman 1958; on Pennsylvania's, see Wilbur 1906; for a general historical discussion, see Dunn et al. 1954, 3, also reprinted in Hetzel 1997.

36. Balfe 1918, 777.

37. On registration's benefits for "commercial progress . . . real estate . . . life insurance," see American Medical Association 1912, 9; and a speech by Metropolitan Life Insurance Company's Louis Dublin (1913).

38. The broader lesson here is that success in practice cannot depend on rational conditions alone. The perspective I adopt holds that rational-conceptual content is just one kind of material contributing to the success of practical ensembles. That said, rational content is a particularly privileged material, for without it we cannot discriminate the difference between purposive practices and systematic automata. See further discussion in Koopman 2017a.

39. Foucault 1990 (orig. pub. 1976).

40. See my further analysis of these themes in the collaboratively authored "Standard Forms of Power" (Critical Genealogies Collaboratory 2018).

41. On black boxes, see Latour 1987, 131.

42. Wilbur 1916, 23.

43. See Hemenway, Davis, and Chapin 1928.

44. Lerrigo 1922, 116.

45. Lerrigo 1922, 117.

46. Lerrigo 1922, 117.

47. Lerrigo 1922, 118.

48. Plecker 1915, 1045.

49. US Census Bureau 1903c, 28.

50. US Census Bureau 1903c, 27.

51. US Census Bureau 1903c, 30.

52. On legacy parish registry systems, see Szreter 2012; on their contemporary significance, see Szreter 2007.

53. US Census Bureau 1903c, 29.

54. US Census Bureau 1903c, 29. On the deficiencies of the book or index system, see US Census Bureau 1903b, 3, 10.

55. US Census Bureau 1903c, 28.

56. This document contained "the Model Law in its original form" (Wilbur 1916, 21). A specification of a standard certificate of birth is offered in pamphlet no. 104 (US Census Bureau 1903c) in §14 of the model law in the form of twenty items of information; a 1907 version of the model law (Wilbur 1916) specifies the standard certificate in §14 in terms of twenty-three required items

of information necessary for "legal, social, and sanitary purposes" (note again the tripartite argumentative structure). A specification of the administrative apparatus for registration is described in §§2–5 and §§16–21 of the 1903 model law printed in pamphlet no. 104 (US Census Bureau 1903c); the 1907 Model Law closely approximates this in its §§2–4 and §§16–20 (Wilbur 1916).

57. US Census Bureau 1906 reprints the Pennsylvania legislation and the text of a speech by Cressy Wilbur delivered there in 1904 to support adoption of the model law.

58. US Census Bureau 1903c, 30.

59. US Census Bureau 1903c, 30.

60. Wilbur 1916, 39. In an earlier publication, Wilbur briefly flirted with a different mood in suggesting that "the persistent worker for an efficient system of vital statistics for the entire United States must necessarily be an optimist," before announcing that he would focus anyway on "the dark side of registration work" (1913, 1253).

61. Chaddock 1911, 72.

62. Chaddock 1911, 63.

63. Chaddock 1911, 67.

64. Chaddock 1911, 66.

65. US Census Bureau 1908, 17.

66. Wilbur 1916, 18.

67. Watters 1924, 105.

68. Chaddock 1911, 70, 71.

69. For a related discussion of social measure, see Mader 2011, 43–70; and 2007.

70. US Census Bureau 1908, 13.

71. US Census Bureau 1908, 14.

72. US Census Bureau 1903c, 41.

73. Pearson 2015, 1147–55.

74. Such strategies of resistance, of course, only make sense against the backdrop of relatively stable powers of identification. This is why, as noted above, I would argue that an analysis of resistance must always follow an analysis of the emergence of stabilizations.

75. For accounts of the Children's Bureau work on registration, see Tobey 1925, 4; US Children's Bureau 1937, 33; Bradbury 1962, 9; and Lindenmeyer 1997, 34.

76. US Children's Bureau (US Department of Health and Human Services) 2012, 32; see discussion by Bradbury 1962, 9; and Lindenmeyer 1997, 23.

77. The Birth Registration Area was implemented in 1916 as applicable to 1915 data sets. For internal evidence of inauguration in the first four months of 1916, see Wilbur (1916, fn1), indicating inauguration subsequent to the delivery of Wilbur's text, which was likely written in late 1915 but not delivered for printing until at least the second week of January 1916 (as suggested by the letter of transmittal included by Census director Sam L. Rogers).

78. Lathrop 1914a, 12.

79. Lathrop noted in her *Second Annual Report* that in 1914 a team of more than 1,500 women volunteers conducted canvasses in seventeen states to produce records of more than 3,400 babies, approximately 700 of whose births were unregistered (1914b, 10). Lathrop's *Third Annual Report* notes 222 committees working in twenty-four states recording 12,865 babies, of whom 3,415 were unregistered (1915, 13). The next year, Lathrop reported tests from 257 cities in twenty-four states (1916, 16).

80. Lathrop 1915, 13.

81. US Children's Bureau 1916, 1919.

82. US Children's Bureau 1919, 12, 9.

83. US Children's Bureau 1916, 6.

84. The Children's Bureau did not so much coordinate a social campaign as they devised a functional technology of information auditing. I thus dissent from Marshall's characterization of the Children's Bureau as having "orchestrated" from 1912 to 1918 "a vigorous national campaign for birth registration . . . conducted mainly at the local level by means of door-to-door investigations" (2012, 460). Campaigns there were, especially in the next decade; but these need to be separated from the specific function of a birth-registration test.

85. US Children's Bureau 1919, 12.

86. Act to Establish Children's Bureau, Sec. 2, 37 *Stat. L.*, 79; see also Tobey 1925, 62.

87. Cf. Lindenmeyer 1997, 15.

88. US Children's Bureau 1937, 2.

89. Marshall 2012, 460; cf. Parker and Carpenter 1981.

90. Parker and Carpenter 1981, 70.

91. Concerning the Census's attempt to contrast their own professed professionalism with an amateurism attributed to the women social workers, it might be thought that that the Children's Bureau were amateurs with respect to their statistical methods, for their method relied on an implicit statistical sampling. According to Lathrop, it was an estimated 5 percent sample size (1914a, 12). The fact that these samples were not controlled for randomization surely led to skewed data. Thus, as soon as the Census Bureau took over the project, they sought methods for increasing the reliability of these data, as later relayed by William H. Davis (1925) against the backdrop of his own earlier survey (W. H. Davis 1917). But what is notable here is that Census would not truly succeed in this effort for at least two more decades, repeating in experiment after experiment many of the initial sampling errors of the Children's Bureau. Not until 1940, seven years after the Birth Registration Area was officially declared complete, was a reliable birth-registration test developed, in the form of a sample of infant births attained by census enumerators (Lenhart 1943, 685). This was also the first year in which the decennial census relied on statistical sampling—a method that the Census Bureau did not adopt into its operations until its 1937 test survey of unemployment. Solving the birth-registration test problem, in other words, seemed to have involved solving the problems to which reliable statistical sampling is a response. This statistical innovation, however, was being worked out by Karl Pearson and others in the years under survey. In other words, the Children's Bureau could not really have been amateurs in a field in which nobody was yet a professional.

92. On the gendered context of opposition to the Children's Bureau's efforts to promote a "maternalist welfare state," see Skocpol 1992, 522–24.

93. For background on Lathrop, see Lindenmeyer 1997, 27–29; Parker and Carpenter 1981, 61; Addams (1935) 2004. For a brief overview of Lathrop's contributions at Hull House that does not mention her later work at the Children's Bureau, see Duran 2014.

94. See, for instance, the contribution to *Hull-House Maps and Papers* by Lathrop ([1895] 2007, 120–29).

95. Theerman 2010, 1590.

96. Lathrop 1917, 1916.

97. US Census Bureau 1917, 9 (this report was published in 1917 with data from 1915); 1918, 19.

98. For one of countless examples of an article by a later researcher at the Children's Bureau itself, see Woodbury 1922, 122.

99. See US Census Beareau 1930a, 1930b, and 1936; these are the bureau's Birth, Stillbirth, and Infant Mortality Statistics reports for 1927, 1928, and 1933, respectively (this is a retitled continuation of the earlier Birth Statistics for the Registration Area of the United States series).

100. See US Census Bureau 1917.

101. American Child Health Association 1920, poster insert.

102. W. II. Davis 1925; for a recent survey history of these campaigns, see Marshall 2012, 460–62.

103. Marshall 2012, 462.

104. For an institutional self-history of the ACHA written at its dissolution, see Van Ingen 1935.

105. Palmer 1933, 132.

106. One such stutter was occasioned by wartime employment exigencies vis-à-vis state administration of delayed registration procedures; see Landrum 2010. Crucial is that this waiver of the birth certificate requirement (by allowing workers to substitute for it the older technology of an affidavit) in this context was only a temporary measure. The waiver's ephemerality proves the rule that the birth certificate had been successfully installed as a core component of American informational identification. Consider an earlier 1941 editorial in the *American Journal of Public Health* noting that "in the past there was not so much need for proof of citizenship, and in such cases as did arise, almost any kind of document or affidavit was accepted as evidence," in contrast to the then-present when "except for a certified copy of birth certificate there has been no commonly accepted basis of proof of citizenship" (*American Journal of Public Health* 1941, 631, 632).

107. W. L. Austin 1935, 597.

108. W. L. Austin 1935, 597.

109. W. L. Austin 1935, 598.

110. Dunn 1935, 1321.

111. See above discussion (pp. 47–48) of Wilbur 1916, 39.

112. Wilbur 1916, 47.

113. Wilbur 1916, 47.

114. See US Social Security Act of 1935, Pub. L. 74-271, 49 Stat. 620, Title VII and Title II (1935).

115. Committee on Economic Security 1935, 45; see also retrospective discussion by the committee's executive director, Edwin Witte (1962, 195–96).

116. Schlesinger 1958, 314.

117. Wyatt and Wandel 1937, 51.

118. On the 1890 US census as the birth of big data, see Monea 2016, 84.

119. Corson 1938, 3; see also Corson as cited in Supervision 1939, 12; and Altmeyer 1966, 87.

120. Committee on Economic Security 1937, 209.

121. Altmeyer 1966, 67.

122. McKinley and Frase 1970, 347.

123. See Wyatt and Wandel 1937, 58–61; Fay and Waserman 1938, 24–25; McKinley and Frase 1970, 368–70; and Cronin 1985, 16–17.

124. See numbers reported by Wyatt and Wandel (1937, 62–63) and McKinley and Frase (1970, 370).

125. According to McKinley and Frase (1970, 364), the third company was Remington Rand Corp., while according to Cronin (1985, 19) it was the Monroe Calculating Machine Co.

126. Commentators at the time thought of the card data as many would today—namely, as a representation of an employee's account. One observer noted of the punch cards that "each hole represents a figure or fact" (Van Boskirk 1939, 21). Another publication suggested that "underlying the whole complicated system . . . is the rather simple technical item that numbers and letters can be represented by means of holes punched through cards" (Wyatt and Wandel 1937, 119). These descriptions sound sensible, but they in fact are mistaken. The cards and holes do not *represent* the account—they *are* the account. For what is an account without an assembly of technical apparatus to access it? In administering Social Security, default authority must go to the information records rather than any contrary claims by a beneficiary. If there is a conflict, it is the record that is assumed to be correct such that the burden of proof (i.e., the burden of adducing contrary records) is on the beneficiary.

127. Altmeyer 1966, 71.

128. McKinley and Frase 1970, 374.

129. Van Boskirk 1939, 20.

130. Discussions of the relationship between birth registration and OAI registration appear in both literatures. In the literature on Social Security, see Wyatt and Wandel 1937, 135; W. C. Smith and Falk 1939, 454; Favinger and Wilcox 1939, 484; and McKinley and Frase 1970, 377. In the birth registration literature, see Deacon 1937, 498; and Reeder 1937, 1216.

131. SSNs are not technologies of identification for all purposes, such as forensic ones. But for purposes of functional bureaucratic identification, SSNs can and do function as identifiers. Technicians at the Social Security Board recognized this from the start: "The first step was that of devising a method of identifying each individual wage earner" (Corson 1938, 3; cf. Wyatt and Wandel 1937, 45).

132. On the life insurance practices informing Social Security, see Wyatt and Wandel (1937, 45, 49, 358). This transfer of techniques is partly explained by the migration of life insurance employees to the Social Security Board, including Isidore S. Falk, who had previously taught Met Life's Louis Dublin at Yale (cf. Bouk 2015, 179), and Bill Williamson, who traveled from Travelers to become chief actuary at the Social Security Board (cf. Bouk 2015, 215). More broadly on information practices in the life insurance industry around the turn of the century, see Yates 2005, 11–109; and Bouk 2015.

133. On the decision to implement the Social Security number, see Wyatt and Wandel 1937; and Corson 1938.

134. McKinley and Frase 1970, 343.

135. McKinley and Frase 1970, 502.

136. McKinley and Frase 1970, 343. Way was not responsible for the original Social Security card itself, which was designed by Frederick E. Happel, an artist and engraver living in Albany, New York (Puckett 2009, 58). Prior to settling on a small wallet-size card as the "identification token" of choice, a variety of other proposals were considered, including one from the Addressograph-Multigraph Corporation of a small metal token .75 inches by 2.875 inches, produced of Monel metal, a noncorrosive nickel-copper composite material, and to be engraved with workers' names and SSNs. McKinley and Frase 1970, 327–29; Altmeyer 1966, 69.

137. For accounts of this spread, see Puckett 2009; see also Social Security Administration, "History FAQ," question 21, accessed September 30, 2017, https://www.ssa.gov/history/hfaq .html; and Social Security Administration, "Social Security Cards: Version History," accessed September 30, 2017, https://www.ssa.gov/history/ssn/ssnversions.html.

138. Syeed and Dexheimer 2017; Newman 2017.

139. DeLillo 1986, 141.

Chapter 2

1. Quoted in Agence France-Presse 2015.

2. These figures are according to reporting by Cadwalladr (2018) and match those on the firm's "CA Political: CA Advantage" webpage at https://ca-political.com/ca-advantage, accessed March 26, 2018.

3. Cadwalladr (2018) and numerous others reported on the use of the Big Five by Cambridge Analytica.

4. See my argument in Koopman 2018.

5. See G. Allport and Odbert 1936.

6. Foucault 1977, 111.

7. Rose 1996, 10.

8. Rose 1996, 10.

9. Rose 1996, 75 (cf. 112). Rose's analysis of the test is indebted to Foucault's discussion of the examination (Foucault 1995 [orig. pub. 1975], 184). My approach distinguishes the broad technology of examinations from more specific technical deployments such as the test in the form of the pencil-and-paper (now computerized) multiple-choice questionnaire. I do so by drawing on Rebecca Lemov's account of a postwar effort to build a comprehensive archive of Micronesian culture. The apparatus this effort deployed included the Rorschach test, the thematic apperception test, and other techniques spawned a few decades earlier by personality researchers. Lemov attends to the techniques of these tests in terms of their data collection methods, informational records, and card storage technologies: "I want to stress the vision and technologies employed rather than the contents" (2009, 60). Drawing on Lemov's analysis, I regard the test as a twentieth-century technique whose specificity differs from the technical specificities that Foucault charted in his analysis of nineteenth-century forms of examination. Foucault's analysis, for instance, is focused on the normalizing yield of the examination, whereas my analysis is focused on the test as a technique of formatting where normalization is either secondary or absent.

10. Rose 1985, 91.

11. Davidson 2001, chap. 1–3.

12. Davidson 1990, 12–21. For a genealogical analysis of the transition from "visible" physicality to "physiological development" in the contexts of sexuality and race, see McWhorter 2009, 121.

13. Krafft-Ebing 1998 (orig. pub. 1886), 53; see discussion in Davidson 2001, 23.

14. Davidson 1990, 24, 13.

15. On these earliest moments of the assembly of personality, see Hacking 1995.

16. Funder 1997, 76.

17. Funder 1997, 60.

18. Larsen and Buss 2010, 3.

19. Rose's Foucauldian description of his method is compelling: "Not a history of ideas, then, but a history of practices, techniques, institutions, and agencies, of the forms of knowledge which made them thinkable and which they, in their turn, transformed. And a history of the categories and problems around which such complex apparatuses formed, which provided the motivation for their emergence and the targets of their tactics" (1985, 6). On the need for connecting practices and concepts in genealogy, see my work in Koopman 2017a building on Davidson 1996, 1997, and 2001. Additional to the contributions of Davidson and Rose, this chapter builds on work by Gould ([1981] 1996), Danziger (1990, 1997), Hacking (1995), Lemov (2005),

McWhorter (2009), and of course Foucault (2006 [orig. pub. 1974], 1995 [orig. pub. 1975], 2003a [orig. pub. 1975], 1990 [orig. pub. 1976]).

20. James (1890) 1950, 1:379–93; see discussion in Leary 1990, 114.

21. James (1890) 1950, 1: 384, 390; see also Hacking 1995, 223.

22. Hacking 1995, 160, 172.

23. See further Hacking 1995, which informs my project both substantively and methodologically. Hacking presents his book as an "archaeology" (4) in Foucault's sense and figures his project as offering a conception of "memoro-politics" (214–15) in parallel to Foucault's conceptions of "bio-politics" and "anatomo-politics."

24. James 1984 (orig. pub. 1895), 320.

25. Dumont 2010, 38; see also Barenbaum and Winter 2008, 4; and Danziger 1997, 125.

26. Susman 1984, 276; see also Haltunnen 1989 and discussion in Nicholson 2003, 3.

27. Danziger 1997, 127.

28. Hacking 1995, 132.

29. On Prince, see Hacking 1995, 44, 132, 231.

30. Prince 1906, 514.

31. See, for instance, Prince 1885, 134.

32. Prince 1914.

33. Prince 1914, 537.

34. Prince 1885, 137.

35. Prince 1914, 109.

36. Prince 1914, 132; Prince refers to the "telegram, Marconigram, and phonogram" (131).

37. See Prince 1914, 3, 117, on these terms, which (along with *reproduction*) are fundamental to his theory.

38. Prince 1914, 254.

39. Prince 1914, 420. Prince had earlier described hypnosis (374) and automatic writing (391) as methods for obtaining "information."

40. Prince does later name and criticize Freud: "It is impossible for a method of this kind to be reliable," he quips of Freud's psychoanalysis, adding that it "falls far short of having that exactness which scientific procedure requires" (1921, 3). For more on Prince and Freud, see Taylor 1928, 101.

41. Prince 1914, 420.

42. Prince 1914, 421, 422, 420; he here also relatedly figures "the content of consciousness" as "information" (344).

43. Freud 1978 (orig. pub. 1926), 85.

44. Galton (1883) 1919.

45. Gould (1981) 1996, 108.

46. Galton 1879, 149.

47. Galton 1884, 179.

48. See Galton (1883) 1919, 17.

49. Galton 1888.

50. Galton 1884, 180.

51. Roback 1932, 214.

52. Woodworth 1933, 74.

53. Woodworth 1933, 74.

54. This has been noted before by other historians of psychology. Rose writes of 1920s psychology as awash in "a flood of literature on tests and testing" (1985, 141); Danziger counts "between

1921 and 1936 some five thousand articles on mental testing" (1990, 223n17); and Carson states that "by the mid 1920s testing itself had become part of the cultural landscape" (2007, 252). Carson further suggests that the early tests from the 1890s and 1900s were "not at all like the modern multiple-choice" tests for intelligence and personality we are still accustomed to today (2007, 141; cf. 173). For a quantitative view, consider the archival collection in the PsycTESTS database maintained by the American Psychological Association. The collection dates to 1896, where the first recorded tests are two association tests by Mary Whiton Calkins. After that, there are no tests from 1897, one from 1898, and between zero and four for each year thereafter until 1917. Then there are nine tests from 1918, seven from 1919, and eighteen from 1920. After a few more years of steady production, there was a significant drop in 1924, after which point the numbers remain relatively low, not picking up again until 1941, when there was a sharp spike to twenty-one tests. Thereafter the number of published psychometric instruments archived at this site steadily increases. The database contains more than 2,039 tests from the year 2013. Records from the Cummings Center for the History of Psychology Archives at the University of Akron indicate a slightly different pattern. These archives suggest steady growth throughout the period, with slight drops from 1932 to 1937 and from 1942 to 1945. My best guess is that some of the fluctuations in these two databases is less representative of real changes in the field and more a function of archiving contingencies.

55. See Franz 1920; May and Hartshorne 1926; Manson 1926; and Roback 1927; these contrast to an earlier but different approach to compiling a volume of tests in Whipple 1910.

56. Galton is often cited as the originator of this specific form of measure, but I find little evidence that Galton sought to test what we would today call "intelligence." Galton was interested both in genius and in the measure of persons, but these two did not seem to link up anywhere in his work into something that might be considered a prototype of today's intelligence tests. See related discussion in Danziger 1997, 66.

57. See Binet and Simon 1906, 1908.

58. See Carson 2007, 156.

59. See Terman 1916.

60. Nicolas et al., 2013, 708. My discussion in this paragraph is indebted on many points to this study and to Rose 1985, 126–131.

61. Whipple 1914–15, 2:674.

62. Whipple 1914–15, 2:674.

63. Whipple 1914–15, 2:689.

64. Franz 1912, 134.

65. See Yerkes 1921; more specifically, in light of the discussion of Binet above, see Yerkes's efforts at distinguishing his committee's psychological work from psychiatry (1921, 88, 98).

66. On the army intelligence tests, see Woodworth 1921, 276.

67. Woodworth 1921, 272, 273.

68. Woodworth 1939 (orig. pub. 1932), 9.

69. Woodworth 1939 (orig. pub. 1932), 12.

70. Woodworth 1921, 285.

71. Woodworth 1921, 286–88.

72. See Gould (1981) 1996, 176; McWhorter 2009, 141.

73. On the test's administration, see Woodworth 1939 (orig. pub. 1932), 19; and Hollingworth 1920, 126.

74. Succinct accounts of the test's development can be found in Franz 1920, 170 (note that Franz's discussion of Woodworth's "Test of Emotional Instability" appears only in the book's

second edition); Hollingworth 1920, 118; Matthews 1923, 2; and Woodworth 1939 (orig. pub. 1932), 18.

75. Franz 1920, 171.

76. Woodworth 1918.

77. Franz 1920, 171.

78. Woodworth 1918.

79. The entire apparatus ought to be eminently familiar to anyone today who has encountered Buzzfeed quizzes purporting to report what spirit animal or which rock star a test taker most resembles.

80. See Hollingworth 1920, 128–46.

81. Woodworth 1939 (orig. pub. 1932), 19.

82. See Matthews 1923.

83. Laird 1925.

84. Papurt 1930, 342–43. Curiously, Papurt proposed to eliminate Woodworth's question 91 with a mocking quip: "Would Woodworth have us believe that the statement, 'It is easy to make me laugh' is a 'pathological' answer?" (349). Interestingly, Franz's reprint of Woodworth's test suggests the opposite valence for the answer: "91. Is it easy to make you laugh? (No)."

85. All these tests are inventoried at the APA PsycTESTS archival database and the University of Akron Cummings Center for the History of Psychology Archives.

86. Barenbaum and Winter 2008, 5.

87. G. Allport 1921, 443; cf. F. Allport and G. Allport 1921, 8.

88. G. Allport 1921, 445.

89. G. Allport 1921, 447.

90. G. Allport 1921, 449.

91. F. Allport and G. Allport 1921, 9.

92. Symonds, 1924.

93. Symonds 1924, 492.

94. Symonds 1924, 492.

95. Symonds 1924, 492–93.

96. See my discussion of these matters in Koopman 2013, 236–38; on measure generally, see Hacking 2009, 96; on social measure specifically, see Mader 2011, 43–70; 2007.

97. I here follow J. L. Austin (1962) 1979, 69.

98. Laird 1925, 420.

99. See Manson 1926.

100. See May and Hartshorne 1926.

101. See Roback 1926.

102. G. Allport 1921, 447.

103. G. Allport 1928, 118; the paragraph in which this sentence appears begins by calling attention to a nominalistic position identified with Symonds 1924 as discussed just above (G. Allport 1928, 118n3).

104. G. Allport 1937, 455.

105. Larsen and Buss 2010, 101.

106. G. Allport 1927, 284, in response to Symonds 1927.

107. G. Allport 1927, 285.

108. G. Allport 1927, 292.

109. See G. Allport 1928 (Allport's report on the test to the field); and G. Allport and F. Allport 1928 (the test itself).

110. Gillespie 2014, 167, 169; Mittelstadt et al. 2016, 1, 14.

111. G. Allport 1928, 122.

112. G. Allport 1928, 136.

113. G. Allport 1928, 126–31.

114. G. Allport and F. Allport 1928, 6.

115. G. Allport and F. Allport 1928, 10.

116. G. Allport and F. Allport 1928, 13.

117. G. Allport and F. Allport 1928, 1.

118. G. Allport 1928, 132.

119. G. Allport 1928, 132.

120. G. Allport 1928, 132.

121. G. Allport 1966, 49.

122. G. Allport and Vernon 1930, 677.

123. G. Allport and Vernon 1930, 702.

124. G. Allport and Vernon 1930, 702.

125. G. Allport and Vernon 1930, 704.

126. G. Allport and Vernon 1930, 706; for their application of internal consistency, see G. Allport and Vernon 1931, 238.

127. G. Allport and Vernon 1930, 707.

128. G. Allport and Vernon 1930, 707.

129. G. Allport and Vernon 1930, 707.

130. G. Allport and Vernon 1930, 712, 713.

131. G. Allport and Vernon 1930, 713.

132. G. Allport 1931.

133. G. Allport 1960, 131.

134. G. Allport 1931, 368.

135. G. Allport 1931, 369; the complete quotation is "the existence of a trait may be established empirically or statistically," but the claim is misleading since the "empirical" is defined as the "casual" approach of case study and biography, whereas "statistical techniques" offer proof "more exactly" (370).

136. Thorndike 1918, 16; quoted in G. Allport and Vernon 1930, 702 (though their citation is for a later formulation).

137. See G. Allport and Vernon 1931, 1933; and G. Allport and Cantril 1935.

138. G. Allport and Odbert 1936.

139. G. Allport and Odbert 1936, 4.

140. G. Allport and Odbert 1936, 20.

141. See G. Allport and Odbert 1936, 22.

142. See G. Allport 1937.

143. G. Allport 1937, vii.

144. G. Allport 1937, vii.

145. Danziger 1990, 163.

146. G. Allport 1937, vii.

147. G. Allport 1937, 48; cf. 46.

148. G. Allport 1937, 48; cf. 46.

149. G. Allport 1937, 52 (italicized in full in the original); for two other uses of this phrasing, cf. G. Allport 1927, 285; and G. Allport and Odbert 1936, vii.

150. G. Allport 1937, 51.

151. G. Allport 1937, 52.

152. G. Allport 1937, 235.

153. G. Allport 1937, 235; the same comparisons are repeated in G. Allport 1958, 112, and G. Allport 1938, 4.

154. G. Allport 1938, 4.

155. G. Allport 1937, ix.

156. G. Allport 1937, 289; this is his characterization of the view described as "the most worthy of endorsement."

157. See G. Allport 1937, 298–300, 312–19.

158. The inaugurating work in contemporary situationism is Mischel 1968.

159. The term *specificists* appears in G. Allport and Odbert 1936, 11, but not in G. Allport 1937.

160. G. Allport 1937, 327.

161. G. Allport 1937, 327, 329.

162. G. Allport 1937, 329.

163. One impressive exemplar is Charles Spearman's effort to mathematically reduce correlations on all intelligence tests down to a single factor, which he named g; see discussion in Gould (1981) 1996, 280.

164. See discussion in Cattell 1957.

165. See the summary by Eysenck and Eysenck (1976).

166. See the summary by Ashton, Lee, and de Vries (2014).

167. G. Allport 1937, 245.

168. G. Allport 1937, 383.

169. G. Allport 1966, 50; and "galloping gamesmanship" (54).

170. G. Allport 1937, 245.

171. G. Allport 1937, 369.

172. G. Allport 1937, 369.

173. See G. Allport 1943.

174. G. Allport 1937, 369.

175. G. Allport 1937, viii, 182, 561.

176. G. Allport 1968a, 384.

177. G. Allport 1968a, 384, citing G. Allport 1953.

178. G. Allport 1953, 102.

179. G. Allport 1953, 103.

180. G. Allport 1937, 187.

181. G. Allport 1937, 187–88.

182. For an account of how a different stream of psychology similarly used behaviorist equations to invade psychoanalytic corners, see Lemov 2005, 131–38; for a discussion of how Wiener sought to turn psychoanalysis into "a process of moving information, not unearthing meaning," see Halpern 2014, 68.

183. G. Allport and Vernon 1930, 689.

184. G. Allport and Vernon 1930, 714.

185. G. Allport 1937, viii.

186. G. Allport 1937, 395, 390.

187. Danziger (1990, 82) charts a similar shift from 1914 to 1949, with a big shift notable between 1924 and 1934—he notes a changing publication strategy at the *Journal of Abnormal and*

Social Psychology which went from 80 percent of its empirical papers being individual case studies in 1924 to only 25 percent in the following year, when it was taken over by the American Psychological Association.

188. G. Allport and Vernon 1930, 716.

189. G. Allport 1937, 399.

190. G. Allport 1937, 390.

191. G. Allport 1937, 391–94.

192. G. Allport 1937, 395.

193. Nicholson 2003, 182; Nicholson suggests that Allport's only case study was the much later *Letters from Jenny* in 1965 (ibid.).

194. He does mention, in a footnote, "an experiment" of his "in teaching by the case method" (G. Allport 1937, 395n40), citing G. Allport 1929. A survey of the title of his publications in the bibliography in G. Allport 1960 suggests that this is his only published discussion of case-study-like materials prior to 1942.

195. See G. Allport 1942.

196. G. Allport 1942, xi.

197. G. Allport 1942, xii, xi (the long passage is in italics in full in the original).

198. G. Allport 1942, xii.

199. G. Allport 1942, xii.

200. G. Allport 1942, xiii.

201. G. Allport 1942, xiv.

202. G. Allport 1942, 16.

203. G. Allport 1942, 18, citing Thomas and Znaniecki 1918–20.

204. G. Allport 1942, 19.

205. See G. Allport 1937, chap. 14.

206. G. Allport 1942, 21.

207. G. Allport 1942, 59.

208. G. Allport 1942, 60.

209. G. Allport 1942, 60.

210. G. Allport 1942, 160.

211. G. Allport 1942, 160; cf. 187.

212. G. Allport 1942, 160.

213. G. Allport 1942, 63.

214. G. Allport 1942, 63. That these hypothetical models may have had their sources in actual studies is possible, owing to an early teaching experiment using this particular text as reported in G. Allport 1929.

215. Foucault 1990 (orig. pub. 1976), 58–65; 1983.

216. G. Allport 1942, 106.

217. Foucault 1990 (orig. pub. 1976), 67, 66; the significance of Foucault's title (*La volonté de savoir*) was lost in translation for the English publication, and should be rendered *The Will to Know* (rather than *An Introduction*).

218. Foucault 1994 (orig. pub. 1966), 559.

219. G. Allport 1937, 370.

220. G. Allport 1942, 167; cf. 172.

221. See Post 1930.

222. Menninger 1930, 21.

223. See Cattell 1950.

224. Cattell 1950, vii, 2.

225. Cattell 1946, 16.

226. On the postwar growth of psychology, including personality psychology, see Herman 1995, 136–42.

227. See G. Allport 1954; see discussion of Allport in recent works in the political theory of race by E. Anderson (2010, 123–27) and Stanley (2017, 121–43).

Chapter 3

1. See Coates 2017 (orig. pub. 2014).

2. On contract selling, see McPherson 1972 and Satter 2009.

3. Coates 2017 (orig. pub. 2014), §7, 194.

4. Coates 2017 (orig. pub. 2014), §1, 171.

5. Coates 2017 (orig. pub. 2014), §2, 172.

6. Coates 2017 (orig. pub. 2014), §10, 206.

7. Jackson 1985, 203.

8. See National Community Reinvestment Coalition, HOLC map, accessed September 13, 2017, http://maps.ncrc.org/holc/. See also the University of Richmond *Mapping Inequality* project, accessed November 16, 2017, https://dsl.richmond.edu/panorama/redlining/; and University of Maryland, T-Races (Testbed for the Redlining Archives of California's Exclusionary Spaces) project, accessed March 13, 2017, http://salt.umd.edu/T-RACES/demo/demo.html.

9. In addition to work arguing for integration cited below, see also recent work on reparations by Balfour (2005), McCarthy (2004), and Zack (2003); and, closest to my own topic here, Kaplan and Valls (2007).

10. See Johnson (1912) 1995.

11. My analysis draws on the metaphor of black-boxing from Latour 1987, 131.

12. I have in view here recent work by E. Anderson (2010), Stanley (2017), and Rothstein (2017).

13. Stanley 2017, 2.

14. E. Anderson 2010, 3.

15. E. Anderson 2010, 194n12, 194n17, 194n18, and later 96, 107; for a similar perspective on pragmatist egalitarianism, see Rondel 2018, 170–78, 188–200.

16. Among the many sources digested by Anderson is Gordon Allport's (1954) psychological study of racial prejudice cited at the end of the previous chapter.

17. Foucault 2011 (orig. pub. 1983), 4.

18. Latour 1986, 277; Foucault is cited in the footnote to this sentence, in which Latour writes of Foucault's focus as that "of micro-powers diffused through the many technologies to discipline and keep in line" (279n18).

19. Bowker and Star 1999, 28.

20. D. Thompson 2016, 6.

21. See D. Thompson 2016, 2n1, citing Bowker and Star 1999 on classification.

22. D. Thompson 2016, 32, 137.

23. Browne 2015, 16.

24. On Foucauldian discipline and beyond, see Browne 2015, 33–42.

25. Browne 2015, 16.

26. French and Browne 2014, 280.

27. Omi and Winant 1986, 64.

28. Though it is odiously present in some of the historical materials I sifted through to write this chapter, I do not here focus on attitudinal-ideational racism. I am rather interested in how the informatics of race can be leveraged into technological racisms such that racisms get built in to technological aspects of social practice. The reason for this focus is that the technologies that bake racism into structuring social practices often outlive the attitudes of their inventors such that the disappearance of overt racist *attitudes* is no guarantee for the disappearance of persisting racist *technological structures*. On the distinction between attitudinal and structural racisms, see canonically Carmichael and Hamilton 1967, 4; and more recently Blum 2002, 24; and Bonilla-Silva 2014, 2.

29. See Noble 2018.

30. See D. Thompson 2016, 245–64; and Williams 2006.

31. Baldwin (1955) 1984, 6.

32. On statistical racism, see Zuberi 2001, 138; on statistical ghettos, see Muhammad 2010, 7.

33. See American Institute of Real Estate Appraisers 1951.

34. NAREB was founded in 1908 as the National Association for Real Estate Exchanges, changed its name to NAREB in 1916, and then later again in 1972 to the National Association of Realtors.

35. Abrams 1955, 164.

36. American Institute of Real Estate Appraisers 1951, 115; cf. 102–3.

37. Foucault 1995 (orig. pub. 1975), 170.

38. On Levittown projects, see Gans (1967) 2017; on contract selling, see McPherson 1972 and Satter 2009; on self-fulfilling prophecies in real estate, see Taeuber and Taeuber 1965, 21.

39. Massey and Denton 1993, 20, table 2.1.

40. Massey and Denton 1993, 19, 24; see, however, the discussion of southern segregation by Du Bois ([1903] 1989, chap. 9), indicating either that residential segregation was more profound than recent sociological statistics suggest, or that it nonetheless felt serious in 1903 to an observer born in 1868 who would of course only see it increase in future decades.

41. Massey and Denton 1993, 47, table 2.3.

42. Massey and Denton 1993, 28–29.

43. Abrams 1955, 209. See a full discussion of the municipal segregation ordinances 1910–1917 by Rice (1968), and two early reports on the Baltimore ordinance by Stephenson (1914) and Hawkins (1911), the latter in the NAACP's *Crisis*.

44. See *Buchanan v. Warley*, 245 U.S. 60 (1917). The challenge in this case was staged by the NAACP to test the constitutionality of racial zoning laws. Buchanan, a white real estate agent, sold property to Warley, then president of the local NAACP branch. Warley refused to make payment, citing the ordinance as preventing his occupation of the premises. Buchanan sued Warley for breach of contract. This forced the question as to whether the ordinances violated requirements of equality in contract rights as set forth by the Fourteenth Amendment. The Supreme Court ruling compelled Warley, the black defendant, to make payment to Buchanan, the white plaintiff, on the grounds that the law on which he relied to deny payment was itself an abrogation of his rights. Warley must abide by the contract and make payment precisely because the terms of the contract could not apply unequally to him and Buchanan. That which made those terms unequal, the Louisville ordinance, was therefore unconstitutional.

45. On the persistence of discriminatory government zoning practices post-*Buchanan*, see Freund 2007, 72–81.

46. See Helper 1969, 224; on racial covenants, see Massey and Denton 1993, 36; and Abrams 1955, 217.

47. Helper 1969, 224 citing Frank Shaw, "The Negro in Harlem," *Real Estate Bulletin* 1, no. 6 (February 1914): 21-23.

48. Helper 1969, 224, 226, quoting committee reports published in the *Chicago Real Estate Board Bulletin* 25, no. 4 (April 1917): 315-17; and no. 11 (November 1917): 623-24.

49. Helper 1969, 225. For statements from the time, see McMichael and Bingham 1923, 182; and a recommendation by NAREB general counsel MacChesney of Chicago's covenant as a national model (1927, 586). MacChesney was the lawyer who helped NAREB secure the term *Realtor* for their exclusive use (1924a, 1924b).

50. Helper 1969, 226-30.

51. See National Association of Real Estate Exchanges 1913.

52. The proposed changes were presented and defended at the meeting by A. H. Barnhisel (see Barnhisel 1924). It is unclear who authored the changes to the code, especially Article 34. Philpott (1978, 190) and Roediger (2006, 171) attribute authorship to MacChesney, but without any citation. I found no original source in NAREB publications evidencing this.

53. NAREB 1924, 64.

54. The covenants and the code were both invalidated by the Supreme Court in their 1948 decision in *Shelley v. Kraemer*, 334 U.S. 1 (1948).

55. Lewis (1922) 1961, 131.

56. Lewis (1922) 1961, 151.

57. This was also the assessment of observers of the time; for instance, Fisher (1930, 10-11).

58. See H. Nelson 1923.

59. See Fisher 1923a.

60. Fisher 1923a, ix.

61. Fisher 1923a, ix.

62. Fisher 1923a, ix.

63. Fisher 1923a, front matter.

64. See Fisher 1923b.

65. See Atkinson 1925.

66. See the discussion of intelligence testing in chap. 2.

67. Atkinson 1925, 23.

68. Atkinson 1925, 24.

69. See T. Nelson 1925.

70. Fisher 1925, 253, citing F. M. Babcock 1924; Zangerle 1924; Hurd 1903; and McMichael and Bingham 1923.

71. F. M. Babcock 1924, front matter.

72. See Stark 1924; and H. Nelson 1925.

73. See Howe 1911.

74. Howe 1911, 99; on the Somers system, see also Pollock and Scholz 1926.

75. Zangerle 1924, 21.

76. Fisher 1923a, 269.

77. Fisher 1923a, 258.

78. Fisher 1923a, 9.

79. McMichael and Bingham 1923, 229.

80. McMichael 1931, 1.

81. Fisher 1930, 7.

82. Fisher 1930, 7.

83. F. M. Babcock 1924, v; and later, "simply the application of the scientific method of analysis" (2).

84. F. M. Babcock 1924, 73.

85. F. M. Babcock 1924, 73–74.

86. F. M. Babcock 1924, 73.

87. F. M. Babcock 1924, 74.

88. F. M. Babcock 1924, 114.

89. I borrow here from Chun's (2016) work on habitual computation.

90. F. M. Babcock 1924, 74, 94, 114.

91. The exception to this is Zangerle 1924, which does not mention race where it would be found in other texts—namely, in discussions of obsolescence (chap. 35); on obsolescence in appraisal, see n. 112, below.

92. See Fisher 1923a; McMichael and Bingham 1923; F. M. Babcock 1924.

93. McMichael and Bingham 1923, 177.

94. McMichael and Bingham 1923, 181.

95. Fisher 1923a, 116.

96. Fisher 1923a, 110.

97. The mixed case of farmland on which owners also lived would present an interesting borderline case, as Fisher himself recognized in an article the next year (1924). There he exhibited numerous quotations from farmer-owners (all attributed to specific individuals but uncited) concerning specific factors in appraisal. Fisher did not comment on these, but his argument clearly relies on them as evidence. What the article thus performed was a stuttering recognition of the need for increased evidence for the sorts of bald assertions characterizing the "common observations" of his 1923 book.

98. F. M. Babcock 1924, 39.

99. F. M. Babcock 1924, 70–71.

100. F. M. Babcock 1924, 74.

101. F. M. Babcock 1924, 56.

102. The six articles included one by Clark in April, one by Blair in July, two by Kniskern (future first president of the AIREA wing of NAREB) in July, one by Stoheker in July, and one by Douglas in November.

103. This article was by Oeland in April.

104. See Nolan 1925b.

105. Bailey 1925, 38, 37.

106. Bailey 1925, 38.

107. Bailey 1925, 38; this feature of Bailey's report cites and borrows from a 1917 proposal titled "A Statistical Study of American Cities," published in *Reed College Record* (though I was not able to locate this document).

108. Nolan 1925a.

109. Nolan 1925a, 40.

110. Nolan 1925a, 40.

111. W. C. Clark 1925, 28.

112. In the technical jargon of the time, *depreciation* referred to loss of value through physical deterioration and *obsolescence* referred to loss of value for any other reason, including presumed social factors.

113. See, for example, Zangerle 1924; Pollock and Scholz 1926; Ruland 1920; and Webb 1925, the last of which notably reproduces a sample appraisal card that does not include blanks for race of building occupants or racial composition of neighborhood (Webb 1925, 39). Earlier, Howe mentions the word *race* only once, to refer to a "horse race" (1911, 113). That race is not explicitated as a factor in these texts does not, of course, mean that these systems had somehow freed themselves of residual racism. The Somers system endorsed by Howe, for instance, left social factors unanalyzed in order to take them at the face value of community opinion in the form of market prices, where they might nevertheless reappear as real factors for market participants that will thereby be implicit in the valuation procedure.

114. See Mertzke 1927, front matter.

115. On declines in attitudinal racism, see Bobo et al. 2012, 47; as the authors point out, a decline in racist attitudes should not be taken to mean that they will soon disappear as if inevitably.

116. H. A. Babcock 1931.

117. Allingham 1931, 200.

118. H. A. Babcock 1931, 263.

119. See F. M. Babcock 1932.

120. F. M. Babcock 1932, 45.

121. See F. M. Babcock 1932, 44–45.

122. F. M. Babcock 1924, 71, 39.

123. F. M. Babcock 1932, 86.

124. F. M. Babcock 1932, 86.

125. F. M. Babcock 1932, 91.

126. F. M. Babcock 1932, 91.

127. See Levy 1931.

128. Gray 1931, 11.

129. Gray 1931, 11.

130. Gray 1931, 11.

131. Gray 1931, 14.

132. Gray 1931, 14.

133. The term *instruction manual* did not begin to circulate until the 1910s and does not really begin its steady increase in usage until the 1930s (see the Google Ngram Viewer at https://books.google.com/ngrams/graph?content=instruction+manual&year_start=1900&corpus=15).

134. See Prouty, Collins, and Prouty 1930; and Reeves 1928.

135. See Boeckh 1936, 1938.

136. Boeckh 1937, 311.

137. McMichael 1931, v.

138. McMichael 1931, 289.

139. McMichael 1931, 290.

140. McMichael 1931, 292.

141. McMichael 1931, 290–91.

142. See McMichael 1931, 17.

143. McMichael 1931, 268.

144. McMichael 1931, 269.

145. McMichael 1931, 271.

146. McMichael 1931, 272.

147. McMichael 1931, 272.

148. McMichael 1931, 277.

149. McMichael 1931, 278.

150. For instance, it adds to the "Blighted Areas" chapter a brief discussion and endorsement of the NAREB plan for neighborhood improvement associations as a means to the technique of restrictive covenants.

151. McMichael 1937, 330.

152. Jackson 1985, 194. On the Federal Home Loan Bank, see Harriss 1951, 8.

153. Harriss 1951, 16. As a percentage of all nonfarm, owner-occupied residences (including those already full paid off), the figure was approximately 20 percent.

154. Harriss 1951, 23.

155. Harriss 1951, 34.

156. Jackson 1985, 197.

157. See Home Owners' Loan Corporation 1933.

158. Home Owners' Loan Corporation 1933, 6.

159. Harriss 1951, 42.

160. Harriss 1951, 43.

161. Harriss 1951, 45.

162. Harriss 1951, 53. Other HOLC forms did not reproduce this racialization so explicitly—for example, loan applications evidently did not ask for applicants' race, although most credit reports obtained on owner-applicants included this data. Harriss 1951, 47.

163. Jackson 1985, 198.

164. See Jackson 1980.

165. See Hillier 2003.

166. Hillier 2003, 415.

167. Jackson 1985, 202.

168. Hillier 2003, 398–402.

169. Hillier 2003, 397.

170. Abrams 1955, 162; Helper 1969, 202.

171. Jackson 1985, 213.

172. See, for example, US Federal Housing Administration 1936b, II.§228 and II.§255; for discussion, see Abrams 1955, 221; and Straus 1952, 222.

173. US Federal Housing Administration 1938, §902.

174. US Federal Housing Administration 1938, §901.

175. US Federal Housing Administration 1938, §929.

176. US Federal Housing Administration 1938, §937; see also US Federal Housing Administration 1936b, II.§233. Both the 1936 and the 1938 *Manual* also explicitly described restrictive covenants as tools for controlling such "racial occupancy" (see US Federal Housing Administration 1936b, II.§228; 1938, §934).

177. See US Federal Housing Administration/Works Progress Administration 1935; the FHA's contribution to the project was coordinated by Ernest Fisher.

178. US Federal Housing Administration 1936a, 41.

179. US Federal Housing Adminstration/Works Progress Administration 1935, §7, 38.

180. Latour 1987, 234; see also Latour 1999, 48.

181. Muhammad 2010, 15.

182. Muhammad 2010, 20.

183. Rosen 1995, 218.

184. Rosen 1995, 219.

185. Foucault 1995 (orig. pub. 1975), 189, 214.

186. Muhammad 2010, 3.

187. See Sellin 1928, 57–58, 60–61 (cited in Muhammad 2010, 2–3); this style of argument is even more clear in Sellin 1935.

188. Sellin 1950, 679.

189. See summary accounts by Bonilla-Silva (2014, 34) and Rothstein (2017, 109–13); more popular coverage by Badger (2015); and scholarship by Rugh and Massey (2010).

190. Caliskan, Bryson, and Narayanan 2017; Noble 2018.

191. Baldwin (1953) 1984, 163.

Chapter 4

1. Wright (1945) 1951, 153–54.

2. On US welfare policy from the Civil War to the New Deal, see Skocpol 1992.

3. Panagia 2019; see on the *dispositif* Foucault 1980 (orig. pub. 1977) and above discussion of the idea of *assembly*, p. 205, n. 36.

4. See the schema on p. 12. These categories are not exhaustively descriptive of power for all possible analytical purposes; see a fuller analysis in Koopman and Matza 2013.

5. Medina 2011, 33.

6. For example, the technology of the automobile (and its correlative techniques of transportation) implemented specific capacities for mobility, and eventually operations of power centered around those capacities (such as an expectation that workers can commute by car).

7. For example, personal automobiles were designed as a technique for the operative transportation of people, and over time helped turn those persons into agents who conceive of themselves as needing automobile transportation such that eventually it became possible to think of oneself as a driver, and in a way that helped further refine automobiles.

8. Foucault and Deleuze 1980 (orig. pub. 1972), 208.

9. Kittler 2010 (orig. pub. 2002), 37.

10. I thus depart from Sterne's "format theory" in its explicit restriction of formats to standards (2012, 22), perhaps coming closer to Tenen's theory of formatting as "tactical abilities to impose structure onto a medium" (2017, 192), though imposition sends the wrong signal.

11. Foucault 2007 (orig. pub. 1978), 56–57.

12. Kittler 2010 (orig. pub. 2002), 37.

13. On other modes of power exhibited by birth registration, see my collaborative work in "Standard Forms of Power" (Critical Genealogies Collaboratory 2018).

14. Numerous moments of documentary and statistical influence abound in Foucault's works from the 1970s. Yet these many references are nowhere in his work drawn together into an explicit recognition of a power of information; in every instance these references are passing asides. Here is a fairly thorough sampling of the role of what I would call "data technology" in Foucault: 2015 (orig. pub. 1973), 91, 131, 193, 235; 2006 (orig. pub. 1974), 49, 55, 158; 2012 (orig. pub. 1974), 150–51; 2003a (orig. pub. 1975), 46; 1995 (orig. pub. 1975), 189, 190, 196, 200, 214, 250, 252; 2003b (orig. pub. 1976), 243, 246; 2007 (orig. pub. 1978), 315. One might also add to this list Foucault's references to statistics as a biopolitical technology (e.g., 1990 [orig. pub. 1976], 25, 144, 146; 2007 [orig. pub. 1978], 29–86, 103–6).

15. For nitpicky scholarly purposes I need to enter here a caveat in light of Foucault having offered, on at least one occasion, a confusing substantive distinction between *biopolitics* and *biopower* in such a way that the latter would include within its operations both *biopolitics* and *anatomopolitics* (1990 [orig. pub. 1976], 139). In spite of Foucault's regrettable nomenclature, I treat *biopolitics* and *biopower* as cognate in the same way that I write of *infopolitics* and *infopower* as cognate labels for a modality of power and the politics organized by that modality.

16. Foucault 2003b (orig. pub. 1976), 257.

17. Foucault 2003b (orig. pub. 1976), 242, 245.

18. Foucault 2003b (orig. pub. 1976), 243, 244, 249.

19. Foucault 1990 (orig. pub. 1976), 139.

20. See, for instance, Galloway 2004, 84–87; Galloway and Thacker 2007, 70–77; Chun 2011b, 122–31; Halpern 2014, 7–8; and, approximating the kind of layered account of power I suggest in the next section, Schüll 2016.

21. On biopolitics and statistics, see Hacking 2015 (orig. pub. 1982), (1990) 2006; and Mader 2011.

22. See the exemplary discussion of "penal tactics" in Foucault 2015 (orig. pub. 1973), 8–12.

23. See arguments that too much identify biopolitics with statistical and calculative technique (thereby neglecting such quintessential biopolitical functions as an enthusiasm for sex) by Galloway (2004, 87), Hull (2013), Halpern (2014, 25), and Cheney-Lippold (2017, 135). For the opposite error of failing to acknowledge any role of statistics in biopolitics at all, see Agamben 1998 (orig. pub. 1995).

24. Nakamura 2002.

25. Lange and Coen 2016.

26. Foucault 2003b (orig. pub. 1976), 243.

27. Foucault 1990 (orig. pub. 1976), 139. This is the other half of the passage quoted above for his characterization of biopower.

28. Foucault 1995 (orig. pub. 1975), 177–83; see p. 228, n. 11, in this volume for Foucault's terminological qualification on normalization and normation.

29. Foucault 1995 (orig. pub. 1975), 170–229; 2003b (orig. pub. 1976), 246.

30. See the overview at the outset of Wood 2007 and more recently in Lyon 2010.

31. I explore this difference in more detail in Koopman 2015a.

32. Foucault 2003a (orig. pub. 1975), 50.

33. Foucault 1990 (orig. pub. 1976), 136.

34. Weber 2004 (orig. pub. 1919), 33; cf. Foucault 1990 (orig. pub. 1976), 135–38.

35. Foucault 1990 (orig. pub. 1976), 83.

36. Foucault 1990 (orig. pub. 1976), 90.

37. Lessig 2006, 1.

38. Chun 2011b, 21–29. Chun (developing ideas from an essay first published in 2004) is arguing against Galloway (2004, 165) and Hayles (2005, 50). More recent contributions to these debates are in Galloway 2012, 69–75 (from a chapter first published in 2006) and Chun 2011a, 99–103.

39. Hu 2015, xiii, xvi; for his complications of the "code is law" formulation, see Hu 2015, 135.

40. Hu 2015, xvi.

41. Hu 2015, xvi–xviii.

42. Perhaps surprisingly, the birth certificates analyzed in chapter 1 played little direct role in exclusionist immigration policies of the time. For example, the 1924 Rogers Act implemented

a nativist quota system on the basis of an administrative apparatus focused on the paperwork of visas and passports; see Zohlberg 1997, 309; and Robertson 2010, 177.

43. Prohibition is surprisingly not among the many qualities Hu ascribes to sovereign power: martiality, materiality, war, centrality, law, territoriality, and decision form his characterization (Hu 2015, xiii, xvi, 82, 93).

44. Kittler 2010 (orig. pub. 2002), 37, discussed above.

45. Hu 2015, xvi, xxix, 96, 139; whereas a graft is completely dependent on that in which it remains forever planted, a true hybrid can subsist on its own once it has been composed from the variety out of which it emerges.

46. Hu 2015, xiii.

47. On strata, see Deleuze 1988 (orig. pub. 1986) commenting on Foucault.

48. Deleuze 1995a (orig. pub. 1990), 1995b (orig. pub. 1990).

49. See Galloway 2004; Lazzarato 2006; and Raley 2009.

50. See Terranova 2004b; Chun 2006, 2011b; and Cheney-Lippold 2017.

51. Chamayou 2013.

52. Stiegler 2013, 25; 2010, 77.

53. Beer 2016.

54. Panagia, n.d.

55. Amoore 2013.

56. Harcourt 2015.

57. Chun 2006, 9. It is worth noting that this paranoid style may be less a function of Deleuze's argument and more that of how his admittedly vague essays on control have been read: as explicitly postbiopolitical. However, as Thomas Nail (2016) has recently shown on the basis of an analysis of Deleuze's (1986) still-untranslated lectures on control power at Vincennes, Deleuze actually offered control power as a restatement of Foucault's arguments about biopower—a point supported by overlapping terminology of control and biopower in both Deleuze 1988 (orig. pub. 1986), 72; and Foucault 1990 (orig. pub. 1976), 139.

58. Deleuze 1995a (orig. pub. 1990), 175.

59. Deleuze 1995b (orig. pub. 1990), 182.

60. Galloway 2004, 3.

61. My understanding of the dangers of paranoid critique discussed in this paragraph is indebted to work forthcoming by Bonnie Sheehey.

62. Harcourt 2015, 14, 232; for a further statement of how he uses Foucault's work in a manner resonant with my approach here, see Harcourt 2013.

63. Harcourt 2015, 1, 3, 14, 163, 232, 249 (also note Harcourt's subtitle, *Desire and Disobedience in the Digital Age*); thus does Harcourt's argument in *Exposed* contrast the dynamics of the digital with the biopolitical logic of actuarial analysis that he believes characterizes the early twentieth century (cf. 163), a history he has charted in more detail in Harcourt 2007.

64. Harcourt 2015, 22.

65. Harcourt 2015, 157; Harcourt's term resonates with the "data double" of Haggerty and Ericson 2000 and the "data-proxy" of G. Smith 2016.

66. Such consistencies are criterial for any distinction between the old and the new. In other words, it cannot be assumed that time is a constant. Rather, the rate at which time passes is itself a variable contingent on different consistencies of practice. It is only when practices operate differently that they can be recognized as new (or old).

67. See Esposito 2008 (orig. pub. 2004), 43; Agamben 1998 (orig. pub. 1995), 6; and more recently, Deutscher 2017, 21–25, 111.

68. See Foucault 1990 (orig. pub. 1976), 107, 138; see also discussion of Foucault's ambiguity on this point by Critical Genealogies Collaboratory 2018, 642.

69. Foucault 2007 (orig. pub. 1978), 8, 107.

70. Erlenbusch-Anderson 2018, 31.

Chapter 5

1. See Dick (1955) 1987.

2. See Wiener (1948) 1961; Shannon 1949 (orig. pub. 1948); and Turing 1950.

3. Dick (1955) 1987, 1.

4. Dick (1955) 1987, 4.

5. Dick (1955) 1987, 4.

6. Dick (1955) 1987, 6.

7. Dick (1955) 1987, 6.

8. Dick (1955) 1987, 20.

9. Du Bois (1903) 1989, 174.

10. Kittler 1999 (orig. pub. 1986), 259; Kittler repeats the claim elsewhere (cf. 2013 [orig. pub. 1990], 180; 2006 [orig. pub. 1996], 183; and 2010 [orig. pub. 2002], 41, 182, 208, 216). See also his argument that the American Civil War was the scene of production for the typewriter (1999 [orig. pub. 1986], 190)—a point also schematically extended across all modern technical history (1999 [orig. pub. 1986], 243). Kittler once went even so far as to claim that "the development of all previous technical media, in the field of computers as well as optical technology, was for purposes directly opposed to cosmic harmony—namely, military purposes" (2010 [orig. pub 2002], 30).

11. Peters writes of information theory as "the child of war" that performs an "intellectual imperialism" (1988, 19, 18); Galison argues that Wiener's cybernetics proposed "a new vision of the world that was to emerge from this secret confluence of war sciences" (1994, 248); and others concur, including Bowker (1993, 117) and Kline (2015, chap. 1). Where there is little emphasis on militarism, there tends to be a focus on that other simplifying boogeyman of the left, capitalism, as in Sterne 2012, 76.

12. Wiener (1948) 1961, 3, quoted in Kittler 1999 (orig. pub. 1986), 259. Kittler was equal parts Foucauldian and Heideggerian, but his suspicions about information theory's origins are wholly indebted to the latter. Despite his clear curiosity about information theory, Kittler's recurring referrals of Shannon and Wiener back to militaristic danger recall the later Heidegger's bristly comments about "the new fundamental science which is called cybernetics": "Cybernetics transforms language into an exchange of news. The arts become regulated-regulating instruments of information. . . . Philosophy is ending in the present age. It has found its place in the scientific attitude of socially active humanity. But the fundamental characteristic of this scientific attitude is its cybernetic, that is, technological character" (Heidegger 1977 [orig. pub. 1966], 376).

13. See Sale and Salisbury 2015, xxiv; Winthrop-Young 2011, 129–43; and Parikka 2012, 95.

14. Keller criticizes as a simplification the claim that the "cybersciences" were "extending the regime of wartime power" (1995, 86). The most detailed historical version of this argument is that by Mindell (2002).

15. Numerous critics have, for example, forcefully argued that postwar information theory relies on a series of heady abstractions—most notably, the mistaken idea that information can

be immaterial. These critics argue that information cannot be immaterial because information is like everything else in operating through specific materialities. Canonical in this vein are the criticisms by Hayles in the field of new media theory (1999, 12–13; 2010, 147) and Peters in the discipline of communications (1999, 169; 2015, 266; 1988, 19). Both of these canonical criticisms were published in 1999, and their materialist arguments prevail in recent contributions by Terranova (2004b, 11), Kirschenbaum (2008, 39), Golumbia (2009, 225), Clarke (2010, 140), Tenen (2017, 9), and others. Such criticisms are positioned as unmasking a set of untenable assumptions on which information theory and other agendas of data culture seem to rest. On my view, a more fruitful critical engagement would come from asking instead how information theory maintains its effectiveness and its robustness despite its reliance on such obvious abstractions. Yes, the model is a massive idealization in the way that it abstracts from the embodiment of human communicators and the material physics of communications systems. But why deny that some abstractions sometimes make themselves practically useful and even possibly practically obligatory? For a model of one such more complicated critical engagement with cybernetics, see Pickering 2010.

16. Zielinski 2006 (orig. pub. 2002), 260.

17. I borrow this term from Livingston (2001, 36), who reactivates it from Brooks (1918).

18. My focus on processes of ordering borrows from Latour 2004 and owes a deep debt to recent pragmatist theories of political order developed by Berk and Galvan (2009) and Jabko and Sheingate (2018).

19. Lest my metaphors mislead, my view is that everything in history is dynamic—time is process all the way down. Cusps are interesting because of the contrast they present between the relatively quick (the transformable) and the imperceptibly slow (the static). But nothing in history is so slow as to be absolutely unmoving, nothing so frozen as to be immune to all heat (for any entropic freeze would be an index of the end of history itself).

20. Eliot 1934, 7.

21. Rosenberg 2013, 36; see further critical analysis in Gregg 2015, 54–56.

22. On information as a "wilderness," see Chaddock 1911, 72; and S. Thompson 1921, 611; our contemporary term *information overload* first entered the popular lexicon with Toffler 1970.

23. In lecture notes from a course given at the University of Chicago in the 1920s, Mead insisted on a "distinction between information and knowledge as discovery through inference" where the latter is conceptualized in such a way as to be "denied to perception as immediate experience" (1938, 54–55). For Mead, the immediately perceptual is the domain of information only, or what many then called "sense-data." On his view, information by itself cannot rise to the level of inferential knowledge. The apotheosis of Mead's distinction was Sellars's masterful *Empiricism and the Philosophy of Mind* ([1956] 1997), in which it is decisively shown that sense data cannot enjoy the kind of independent epistemic function attributed to them by seventeenth- and eighteenth-century British empiricist epistemology. Mead and Sellars stood in striking contrast to contemporaries trying to save the basic structure of eighteenth-century empiricist epistemology by presenting unanalyzable data as a given foundation for knowledge; see Carnap 1967 (orig. pub. 1928). For Sellars, sense data qua data are devoid of any *independent* epistemic and semantic status. His argument is that data cannot independently carry inferential relations to other cognitive contents, for the reason that inferential relations depend on shared meaning and knowledge.

24. Foucault 1988; translation from the original French (though first published in English) modified.

25. Foucault 1978, 236.

26. Harcourt 2015, 261.

27. Harcourt 2015, 261.

28. My argument does not depend on it, but I think there is a neglected conceptual link between the rise of biopolitics and the rise of communicative democracy. Communicative democracy is not designed so much for the levied subject of sovereignty as it is for the subjects of biopolitics whose lives are invested in food nutrition labels, the birth control pill, and the fiscal policies enabling them. Consider Dewey's claim that "the questions of most concern at present may be said to be matters like sanitation, public health, healthful and adequate housing, transportation, planning of cities, regulation and distribution of immigrants" ([1927] 2016, 124). These are all paradigmatically biopolitical matters.

29. Peters 1999, 10. According to Peters, the modern form of the problematic of communication was first isolated in the 1880s and 1890s (1999, 9–10), which is roughly the same as Nunberg's dating of the emergence of information as a mass noun—that is, as a term that refers to a mass of facts, records, or data (Nunberg 2012).

30. See Shannon 1949 (orig. pub. 1948).

31. Shannon quoted in von Foerster 1952, 207. In his 1956 article "The Bandwagon," Shannon cautioned those who sought rapid deployment of information theory beyond his intended use of "a technical tool for the communication engineer" (Shannon 1956, 3). Shannon's repeated emphasis on this distinction is missing in much critical literature on information theory. It has, however, recently been observed by Kline, who notes Shannon's efforts in "boundary work" separating his communication theory from a more general information theory (Kline 2015, 126). Kline suggests that Shannon failed to hold the line—a point evidenced by early introductory texts on information theory. Little more than a decade after Shannon's original article, Pierce ([1961] 1980, 1; cf. vii and ix from the 1979 preface), Reza (1961, ix), and Ash ([1965] 1990, vi) all explicitly used "information theory" as a synonym for either "communication theory" or "theory of communication," though it is clear that, for them, both terms refer to a theory of efficient encoding for communications systems, and not anything like a complete theory of information itself. These 1960s texts contrast with Weaver's popularizing account of Shannon's theory from 1949 in which it is abundantly clear from the first sentence that the primary concern of the theory is with communication itself (Weaver 1949b, 95).

32. See Wiener (1948) 1961.

33. Wiener (1948) 1961, 156.

34. Wiener (1950) 1988, 27, 16.

35. Wiener (1950) 1988, 17.

36. This sentence is my revision of a statement I initially came across in the work of philosopher of information Luciano Floridi. His sentence, which he presents as a quotation from Warren Weaver, is as follows: "The mathematical theory of communication deals with the carriers of information, symbols and signals, not with information itself" (2010, 45; this quotation appears in many places in Floridi's work, including his *Stanford Encyclopedia of Philosophy* entry "Semantic Conceptions of Information," last accessed December 29, 2018). Floridi cites this sentence as part of a longer quotation from an article by Warren Weaver, an early popularizer of Shannon's mathematical theory of communication. However, the sentence does not appear at the location of Floridi's citation (which is Weaver 1949a, 12) or anywhere else in Weaver's two famous publications on Shannon from 1949; that is, the short version in *Scientific American* (Weaver 1949a) and the long version reprinted in book form alongside Shannon's original article (Weaver 1949b). In preparing my manuscript, I was initially taken by the idea that Weaver's argument

anticipated my own so directly. Since that turns out to be not as obviously true as I had thought, I have elected to crib from Floridi his apocryphal Weaver sentence. This noted, it is amply clear that Weaver does in fact anticipate my broader point that information theory is above all a theory of communication—for instance, in his claims that Shannon's theory addresses "the real inner core of the communication problem" and constitutes "an important contribution . . . to any possible general theory of communication" (Weaver 1949b, 115).

37. Halpern 2014, 74.

38. McKeon 1990 (orig. pub. 1957), 89.

39. McKeon 1990 (orig. pub. 1957), 96.

40. See Habermas 1985 (orig. pub. 1981); and Rawls 1971.

41. See Dewey (1927) 2016; and Lippmann (1922) 1997.

42. Chambers 2003, 308.

43. Most prominent are criticisms of communicative theories of democracy as unwilling to countenance the political work of affects and emotions; see Connolly 1999; Young 2000; and Krause 2008. More poignant for my argument are claims for a conception of communicative public spheres not restricted solely to discourse but also encompassing practical and material conditions of communication; on this, also taking a genealogical perspective close to my own, see Olson 2016.

44. I know of no sustained discussion of the resonance between Shannon's information theory and Habermas's communicative rationality. My account is here an alternative to Geoghegan's (2011) suggestion that American information theory found its European counterpart in French poststructuralism.

45. Sterne 2012, 88.

46. Habermas 1996 (orig. pub. 1992), 296.

47. Habermas 1996 (orig. pub. 1992), 107; earlier: "Every valid norm would meet with the approval of all concerned if they could take part in a practical discourse" (1990 [orig. pub. 1983], 121).

48. Habermas 1985 (orig. pub. 1981), 2:95.

49. Habermas 1996 (orig. pub. 1992), 108.

50. It might be replied that Habermas could in fact countenance formats as instances of strategic action in systems contexts; this, however, underscores my point that Habermas's perspective of communicative action is itself unable to countenance the normative dimensions of information as such.

51. This point clarifies my resonance with Panagia's analysis of Foucault's *dispositif* as "not an instrument of meaning-transmission on the model of a linguistic utterance" but rather an account of "dispositional powers" (Panagia 2019). My view similarly locates a *dispositif* of information that is irreducible to how information can be used for the communication of meanings.

52. See my discussion in Koopman 2018.

53. Moore 2017.

54. See Syeed and Dexheimer 2017; and Newman 2017.

55. Woolley 2017.

56. Habermas was never as explicit as his predecessors; but for early remarks expressing caution about, if not contempt for, information, see Habermas 1991 (orig. pub. 1962), 36, 75, 194, 203.

57. Horkheimer and Adorno (1944) 2002, xvii; cf. 163.

58. Benjamin (1936) 1968, 89.

59. Benjamin (1936) 1968, 89.

60. On Dewey as a precedent for Habermas's own deliberative democracy, see Habermas 1996 (orig. pub. 1992), 304, 316.

61. Dewey (1927) 2016, 225. If Benjamin and Adorno are direct precedents for Habermas, then Dewey played that role for Habermas's contemporary Richard Rorty (1989) and his "conversational" conception of liberal democracy.

62. Dewey (1927) 2016, 225.

63. Dewey (1927) 2016, 170; cf. 204, 227.

64. Dewey (1927) 2016, 149.

65. Wiener (1948) 1961, 95; (1950) 1988, 127.

66. Wiener (1948) 1961, 160; cf. Dewey (1927) 2016, 88.

67. Wiener (1948) 1961, 160, 161; cf. Dewey (1927) 2016, 94, 157.

68. Wiener (1948) 1961, 161; cf. Dewey (1927) 2016, 170, 233.

69. Dewey (1927) 2016, 134; cf. Wiener (1950) 1988, 131. On Wiener's cybernetics and pragmatist communication, see a recent discussion by Peters and Peters (2010).

70. Dewey (1927) 2016, 191.

71. Misak 2017, 297 (cf. 2013, 135). Though Misak endorses the quoted aspect of Dewey's view, she does not endorse the full position; for a fuller analysis of Deweyan deliberative democracy, see Rogers 2009, 210–12.

72. Lippmann (1922) 1997, (1925) 1993.

73. Lippmann (1922) 1997, 171. For Dewey's praise of Jefferson, see Dewey (1940) 1963.

74. Dewey (1927) 2016, 211.

75. Lippmann (1922) 1997, 242, 197.

76. Lippmann (1922) 1997, 246, 245, 239.

77. Lippmann (1922) 1997, 233, 234.

78. Lippmann (1922) 1997, 241.

79. Lippmann (1922) 1997, 241; for insightful discussion of Lippmann's contribution, see Marres 2012, 28–59.

80. Lippmann (1922) 1997, 255–56. To be clear, Lippmann's view (and mine) is that both communication and information are crucial sites for democratic politics. My argument, which I take Lippmann to have anticipated, is against communicativism, or the idea that what Lippmann calls "the machinery of human communication" ([1922] 1997, 8) is the *only* crucial site of politics.

81. On Lippmann as an elitist, see Carey 2009 (orig. pub. 1982), 57; and Peters 1989, 207–11; more recently, alternative interpretations have been offered in Schudson 2008 and Jansen 2009.

82. Lippmann's reputation as assuming elitism in his 1922 book is unjustified. Certainly his later work, as represented in Lippmann 1955, appears rather elitist. But these later writings contrast sharply to Lippmann's book on public opinion ([1922] 1997) and the progressivism of his first two books ([1913] 1969, [1914] 1985). If one had to draw a single line dividing Lippmann's early democratic experimentalism from his later antidemocratic attitude, it should be set between chapters 3 and 4 of *The Phantom Public* ([1925] 1993). Chapters 1 through 3 restate the *Public Opinion* argument that modern democracy suffers from epistemic deficits. Chapter 4 then launches an entirely separate argument about a participation deficit in democracy. Lippmann took these two arguments to be related, yet he offered no justification that they are. In fact, they are unrelated. As I read him, Lippmann's 1922 proposal for a remediation of epistemic deficits was abandoned in 1925 because of his pessimism about irremediable democratic participation. Further in support of locating the dividing line somewhere after Lippmann had written *Public*

Opinion is a six-part series of articles from late 1922 attacking Lewis Terman's then-new intelligence tests (discussed above in chapter 2). Lippmann was relentless against hereditarian proposals that such tests could measure inborn intelligence, which he mocked as "a method of stamping a permanent sense of inferiority upon the soul of a child" (1922, 297). Lippmann's distinction in *Public Opinion* between "insiders" and "outsiders" ([1922] 1997, 251) was thus not one of status or birthright but only a distinction in amount of prior education in a matter of public affairs.

83. Deleuze and Guattari 1994 (orig. pub. 1991), 108.

84. I should also note that I do not construe the technical option as a return to an analysis of materialities, though this is a common association in the literature. Materialism, even in its most vibrant forms, is inevitably a category of metaphysical analysis. Techniques and technologies, by contrast, are categories for the analysis of practice, conduct, and action.

85. I offer such an analysis of James in Koopman 2016.

86. See Sandvig et al. 2014; Lessig 2004; Poitras et al. 2015; Wark 2004; Singer 2011; Brunton and Nissenbaum 2015; and Gehl 2018; I am aware that this is a disparate collection.

Bibliography

Abbate, Janet. 1999. *Inventing the Internet*. Cambridge: MIT Press.

Abrams, Charles. 1955. *Forbidden Neighbors: A Study of Prejudice in Housing*. New York: Harper & Brothers.

Addams, Jane. (1935) 2004. *My Friend, Julia Lathrop*. Urbana: University of Illinois Press.

Agamben, Giorgio. 1998. *Homo Sacer: Sovereign Power and Bare Life*. Translated by D. Heller-Roazen. Stanford: Stanford University Press. Originally published as *Homo sacer: Il potere sovrano e la nuda vita*, 1995.

Agar, Jon. 2001. "Modern Horrors: British Identity and Identity Cards." In Caplan and Torpey 2001, 101–20.

———. 2003. *The Government Machine: A Revolutionary History of the Computer*. Cambridge: MIT Press.

Agence France-Presse. 2015. "'Humans May Soon Download Their Personalities on Computers.'" *Times of India*, April 4. Available at https://web.archive.org/web/20160828171141 /http://timesofindia.indiatimes.com//tech/tech-news/Humans-may-soovn-download -their-personalities-on-computers/articleshow/46805544.cms.

American Institute of Real Estate Appraisers, Education Committee. 1951. *The Appraisal of Real Estate*. Chicago: AIREA.

Allingham, A. P. 1931. "Demonstration Appraisal of a Residence." In H. Babcock 1931, 183–202.

Allport, Floyd, and Gordon Allport. 1921. "Personality Traits: Their Classification and Measurement." *Journal of Abnormal Psychology and Social Psychology* 16, no. 1 (April–June): 6–40.

Allport, Gordon. 1921. "Personality and Character." *Psychological Bulletin* 18, no. 9 (September): 441–55.

———. 1927. "Concepts of Traits and Personality." *Psychological Bulletin* 24, no. 5 (May): 284–93.

———. 1928. "A Test for Ascendance-Submission." *Journal of Abnormal and Social Psychology* 23, no. 2 (July): 118–36.

———. 1929. "The Study of Personality by the Intuitive Method: An Experiment in Teaching from *The Locomotive God*." *Journal of Abnormal and Social Psychology* 24, no. 1: (April): 14–27.

———. 1931. "What Is a Trait of Personality?" *Journal of Abnormal and Social Psychology* 25, no. 4 (January): 368–72.

————. 1937. *Personality: A Psychological Interpretation*. New York: Henry Holt.

————. 1938. "Personality: A Problem for Science or for Art?" In G. Allport 1960, 3–16.

————. 1942. *The Use of Personal Documents in Psychological Science*. Monograph no. 49. New York: Social Science Research Council.

————. 1943. "The Productive Paradoxes of William James." In G. Allport 1968b, 298–325.

————. 1953. "The Trend in Motivational Theory." In G. Allport 1960, 95–110.

————. 1954. *The Nature of Prejudice*. Cambridge: Addison-Wesley.

————. 1958. "What Units Shall We Employ?" In G. Allport 1960, 111–30.

————. 1960. *Personality and Social Encounter: Selected Essays*. Boston: Beacon.

————. 1966. "Traits Revisited." In G. Allport 1968b, 43–66.

————. 1968a. "An Autobiography." In G. Allport 1968b, 376–409.

————. 1968b. *The Person in Psychology: Selected Essays*. Boston: Beacon.

Allport, Gordon, and Floyd Henry Allport. 1928. *The A-S Reaction Study: A Scale for Measuring Ascendance-Submission in Personality: Manual for Directions, Scoring Values, and Norms*. Boston: Houghton Mifflin, 1928 [also reprinted in a 1939 edition that is by far the easier to locate].

Allport, Gordon, and Hadley Cantril. 1935. *The Psychology of the Radio*. New York: Harper and Brothers.

Allport, Gordon, and H. S. Odbert. 1936. "Trait-Names: A Psycho-lexical Study." *Psychological Monographs* 47, no. 1: i–171.

Allport, Gordon, and Philip Vernon. 1930. "The Field of Personality." *Psychological Bulletin* 27, no. 10 (December): 677–730.

————. 1931. "A Test for Personal Values." *Journal of Abnormal and Social Psychology* 26, no. 3 (October–December): 231–48.

————. 1933. *Studies in Expressive Movement*. New York: Macmillan.

Altmeyer, Arthur J. 1966. *The Formative Years of Social Security*. Madison: University of Wisconsin Press.

American Child Health Association. 1920. *Statistical Report of Infant Mortality: In 519 Cities of the United States*. Baltimore: ACHA.

————. 1927. *Five Years of the American Child Health Association: A Bird's-Eye View*. New York: ACHA.

————. 1929. *Teamwork for Child Health: Suggestions for the Year-Round Program in Home, School and Community, How May Day Can Help*. New York: ACHA.

American Journal of Public Health. 1941. "Delayed Birth Registration." *American Journal of Public Health and the Nation's Health* 31, no. 6 (June): 631–33.

American Medical Association. 1912. *Why Should Births and Deaths Be Registered?* Chicago: American Medical Association.

————. 1915. "The Bookkeeping of Humanity." *Journal of the American Medical Association* 65, no. 2 (January 1915): 154–55.

Amoore, Louise. 2008. "Governing by Identity." In *Playing the Identity Card: Surveillance, Security, and Identification in Global Perspective*, edited by C. Bennett and D. Lyon, 21–36. New York: Routledge.

————. 2013. *The Politics of Possibility: Risk and Security Beyond Probability*. Durham: Duke University Press.

Anderson, Elizabeth. 2010. *The Imperative of Integration*. Princeton: Princeton University Press.

Anderson, Sherwood. (1919) 1995. *Winesburg, Ohio*. New York: Modern Library.

Ash, Robert. (1965) 1990. *Information Theory*. New York: Dover.

Ashton, M. C., K. Lee, and R. E. de Vries. 2014. "The HEXACO Honesty-Humility, Agreeableness, and Emotionality Factors: A Review of Research and Theory." *Personality and Social Psychology Review* 18, no. 2 (February): 139–52.

Atkinson, Henry Grant. 1925. "Standard Test for Real Estate." *National Real Estate Journal* 26, no. 16 (August 10): 23–25.

Austin, John L. (1962) 1979. *Sense and Sensibilia*. London: Oxford University Press.

Austin, W. L. 1935. "Vital Statistics in the Census Bureau." *Journal of the American Statistical Association* 30, 191A: 597–99.

Babcock, Frederick Morrison. 1924. *The Appraisal of Real Estate*. New York: Macmillan.

———. 1932. *The Valuation of Real Estate*. New York: McGraw-Hill.

Babcock, Henry A., ed. 1931. *Real Estate Appraisals: Discussions and Examples of Current Technique*. Chicago: National Association of Real Estate Boards.

Badger, Emily. 2015. "Redlining: Still a Thing." *Washington Post*, May 28. https://www.washingtonpost.com/news/wonk/wp/2015/05/28/evidence-that-banks-still-deny-black-borrowers-just-as-they-did-50-years-ago/.

Bailey, William. 1925. "Appraising Your City." *National Real Estate Journal* 26, no. 10 (May 18): 37–39.

Baldwin, James. (1953) 1984. "Stranger in the Village." In *Notes of a Native Son*, 159–75. Boston: Beacon Press.

———. (1955) 1984. "Autobiographical Notes." In *Notes of a Native Son*, 3–11. Boston: Beacon Press.

———. (1972) 2007. *No Name in the Street*. New York: Vintage.

Balfe, James P. 1918. "Birth Registration in Connecticut." *American Journal of Public Health* 8, no. 10 (October): 776–79.

Balfour, Lawrie. 2005. "Reparations after Identity Politics." *Political Theory* 33, no. 6 (December): 786–811.

Barenbaum, Nicole B., and David G. Winter. 2008. "History of Modern Personality Theory and Research." In *Handbook of Personality: Theory and Research*, edited by O. John, R. Robins, and L. Pervin, 3–26. New York: Guilford Press.

Barnhisel, A. H. 1924. "What a Code of Ethics for All Realtors Should Contain." In *Annals of Real Estate Practice*, 39–45. Chicago: National Association of Real Estate Brokers.

Bassett, Caroline. 2015. "Staring into the Sun." In Sale and Salisbury 2015, 178–93.

Beer, David. 2016. *Metric Power*. London: Palgrave Macmillan.

Beniger, James. 1989. *The Control Revolution: Technological and Economic Origins of the Information Society*. Cambridge: Harvard University Press.

Benjamin, Walter. (1936) 1968. "The Storyteller." In *Illuminations*, 83–110. New York: Schocken Books.

Bentham, Jeremy. 1838. *Principles of Penal Law*. In *Works of Jeremy Bentham, Volume One*. Edited by J. Bowring. Edinburgh: William Tate.

Berk, Gerald, and Dennis Galvan. 2009. "How People Experience and Change Institutions: a Field Guide to Creative Syncretism." *Theory and Society* 38, no. 6: 543–80.

Binet, Alfred, and Théodore Simon. 1906. "Méthodes nouvelles pour diagnostiquer l'idiotie, l'imbécillité et la débilité mentale." In *Atti del V congresso internazionale di psicologia*, 507–9. Rome: Forzani.

———. 1908. "Le développement de l'intelligence chez les enfants." *L'Année Psychologique* 14, no. 1. 1–94.

Blum, Lawrence. 2002. *"I'm Not a Racist, But . . .": The Moral Quandary of Race*. Ithaca: Cornell University Press.

Bobo, Lawrence D., Camille Z. Charles, Maria Krysan, and Alicia D. Simmons. 2012. "The Real Record on Racial Attitudes." In *Social Trends in American Life*, edited by P. Marsden, 38–83. Princeton: Princeton University Press.

Boeckh, E. H. 1937. *Boeckh's Manual of Appraisals*, 3rd ed. Indianapolis: Rough Notes.

———. 1936. *Boeckh Index Calculator: For Computing Boeckh Building Cost Local Index Numbers*. Indianapolis: Rough Notes.

———. 1938. *Boeckh Index Calculator Tables: For Computing Boeckh Building Cost Local Index Numbers*. Indianapolis: Rough Notes.

Bonilla-Silva, Eduardo. 2014. *Racism without Racists*. Lanham: Rowman and Littlefield.

Bouk, Dan. 2015. *How Our Days Became Numbered: Risk and the Rise of the Statistical Individual*. Chicago: University of Chicago Press.

Bowker, Geoffrey C. 1993. "How to Be Universal: Some Cybernetic Strategies, 1943–70." *Social Studies of Science* 23, no. 1 (February): 107–27.

Bowker, Geoffrey C., and Susan Leigh Star. 1999. *Sorting Things Out: Classification and Its Consequences*. Cambridge: MIT Press.

Bradbury, Dorothy E. 1962. *Four Decades of Action for Children: A Short History of the Children's Bureau*. Washington: Government Printing Office.

Breckenridge, Keith, and Simon Szreter, eds. 2012. *Registration and Recognition: Documenting the Person in World History*. Oxford: Oxford University Press (for the British Academy).

Brooks, Van Wyck. 1918. "On Creating a Usable Past." *Dial* 64, no. 764 (April 11): 337–41.

Browne, Simone. 2015. *Dark Matters: On the Surveillance of Blackness*. Durham: Duke University Press.

Brunton, Finn, and Helen Nissenbaum. 2015. *Obfuscation: A User's Guide for Privacy and Protest*. Cambridge: MIT Press.

Cadwalladr, Carole. 2018. "'I Created Steve Bannon's Psychological Warfare Tool': Meet the Data War Whistleblower." *Guardian*, March 17. Available at https://web.archive.org/web/20180317181454/https://www.theguardian.com/news/2018/mar/17/data-war-whistleblower-christopher-wylie-faceook-nix-bannon-trump.

Caliskan, Aylin, Joanna J. Bryson, and Arvind Narayanan. 2017. "Semantics Derived Automatically from Language Corpora Contain Human-Like Biases." *Science* 356, no. 6334 (April 14): 183–86.

Caplan, Jane. 2001. "'This or That Particular Person': Protocols of Identification in Nineteenth-Century Europe." In Caplan and Torpey 2001, 49–67.

Caplan, Jane, and John Torpey, eds. 2001. *Documenting Individual Identity: The Development of State Practices in the Modern World*. Princeton: Princeton University Press.

Carey, James W. 2009. "Reconceiving 'Mass' and 'Media.'" In *Communication as Culture*, rev. ed., 53–67. New York: Routledge. Originally published in 1982.

Carmichael, Stokely, and Charles V. Hamilton. 1967. *Black Power: The Politics of Liberation in America*. New York: Vintage.

Carnap, Rudolf. 1967. *The Logical Structure of the World*. Translated by R. George. Berkeley: University of California Press. Originally published as *Der logische Aufbau der Welt*, 1928.

Carson, John. 2007. *The Measure of Merit: Talents, Intelligence, and Inequality in the French and American Republics, 1750–1940*. Princeton: Princeton University Press.

Cartwright, Julyan H. E., Simone Giannerini, and Diego L. González, eds. 2016. "Theme Issue 'DNA as Information.'" *Philosophical Transactions of the Royal Society A* 374, no. 2063 (March). https://rsta.royalsocietypublishing.org/content/374/2063.

Cartwright, Nancy, Jordi Cat, Lola Fleck, and Thomas E. Uebel. 1996. *Otto Neurath: Philosophy Between Science and Politics*. Cambridge: Cambridge University Press.

Cattell, Raymond B. 1946. *Description and Measurement of Personality*. New York: World Book.

———. 1950. *Personality: A Systematic Theoretical and Factual Study*. New York: McGraw-Hill.

Cattell, Raymond B. 1957. *Personality and Motivation Structure and Measurement*. New York: World Book.

Chaddock, Robert Emmet. 1911. "Sources of Information upon the Public Health Movement." *Annals of the American Academy of Political and Social Science* 37, no. 2 (March): 61–76.

Chamayou, Grégoire. 2013. "Fichte's Passport: A Philosophy of the Police." Translated by K. Aarons. *Theory & Event* 16, no. 2. https://muse.jhu.edu/article/509902.

Chambers, Simone. 2003. "Deliberative Democratic Theory." *Annual Review of Political Science* 6:307–26.

Cheney-Lippold, John. 2017. *We Are Data: Algorithms and the Making of Our Digital Selves*. New York: New York University Press.

Chun, Wendy Hui Kyong. 2006. *Control and Freedom: Power and Paranoia in the Age of Fiber Optics*. Cambridge: MIT Press.

———. 2011a. "Crisis, Crisis, Crisis, or Sovereignty and Networks." *Theory, Culture & Society* 28, no. 6: 91–112.

———. 2011b. *Programmed Visions: Software and Memory*. Cambridge: MIT Press.

———. 2016. *Updating to Remain the Same: Habitual New Media*. Cambridge: MIT Press.

Clark, Andy, and David Chalmers. 1998. "The Extended Mind." *Analysis* 58, no. 1 (January): 10–23.

Clark, W. C. 1925. "Causes of Obsolescence." *National Real Estate Journal* 26, no. 18 (September 7): 26–29.

Clarke, Bruce. 2010. "Communication." In Mitchell and Hansen 2010, 131–44.

Coates, Ta-Nehisi. 2017. "The Case for Reparations." In *We Were Eight Years in Power: An American Tragedy*, 163–208. New York: Penguin Random House. Originally published in 2014.

Cole, Simon. 2001. *Suspect Identities: A History of Fingerprinting and Criminal Identification*. Cambridge: Harvard University Press.

Committee on Economic Security. 1935. *Report to the President of the Committee on Economic Security*. Washington: Government Printing Office.

———. 1937. *Social Security in America: The Factual Background of the Social Security Act as Summarized from Staff Reports to the Committee on Economic Security*. Washington: Government Printing Office.

Commonwealth of Pennsylvania. 1852. *Laws of the General Assembly of the Commonwealth of Pennsylvania Passed at the Session of 1852*. Harrisburg: Theo. Fenn.

Connolly, William. 1999. *Why I Am Not a Secularist*. Minneapolis: University of Minnesota Press.

Cook, Jia-Rui. 2017. "Why You Don't Want to Have a Baby in a Car." *New York Times*, June 27. https://www.nytimes.com/2017/06/27/well/family/why-you-dont-want-to-have-a-baby-in-a-car.html.

Corson, John J. 1938. "Administering Old-Age Insurance." in *Social Security Bulletin* 1, no. 5 (May): 3–6.

Critical Genealogies Collaboratory (coauthored by Colin Koopman, Bonnie Sheehey, Patrick Jones, Laura Smithers, Sarah Hamid, and Claire Pickard). 2018. "Standard Forms of Power: Biopower and Sovereign Power in the Technology of the US Birth Certificate, 1903-1935." *Constellations* 25, no. 4 (December): 641–56.

Cronin, Michael A. 1985. "Fifty Years of Operations in the Social Security Administration." *Social Security Bulletin* 48, no. 6 (June): 14–26.

Currah, Paisley, and Lisa Jean Moore. 2009. "'We Won't Know Who You Are': Contesting Sex Designations in New York City Birth Certificates." *Hypatia* 24, no. 3 (Summer): 113–134.

———. 2015. "Legally Sexed: Birth Certificates and Transgender Citizens." *Feminist Surveillance Studies*, edited by R. Dubrofsky and S. Magnet, 58–76. Durham: Duke University Press.

Danziger, Kurt. 1990. *Constructing the Subject: Historical Origins of Psychological Research*. Cambridge: Cambridge University Press.

———. 1997. *Naming the Mind: How Psychology Found Its Language*. Thousand Oaks: Sage.

Davidson, Arnold. 1990. "Closing Up the Corpses." In Davidson 2001, 1–29.

———. 1996. "Styles of Reasoning, Conceptual History, and the Emergence of Psychiatry." In *The Disunity of Science*, edited by P. Galison and D. Stump, 75–10. Stanford: Stanford University Press.

———. 1997. "Structures and Strategies of Discourse. Remarks Towards a History of Foucault's Philosophy of Language." In *Foucault and His Interlocutors*, edited by A. Davidson, 1–17. Chicago: University of Chicago Press.

———. 2001. *The Emergence of Sexuality: Historical Epistemology and the Formation of Concepts*. Cambridge: Harvard University Press.

Davis, A. T. 1926. "Aids in Securing Better Registration." *American Journal of Public Health* 16, no. 1 (January): 25–29.

Davis, William H. 1917. "A Check for the Registration of Births." *American Journal of Public Health* 7, no. 9 (September): 762–64.

———. 1925. "Necessity for Completing the Registration Area by 1930." *American Journal of Public Health* 15, no. 5 (May): 399–404.

Deacon, W. J. W. 1937. "Test and Promotion of Registration of Births and Deaths." *American Journal of Public Health and the Nation's Health* 27, no. 5 (May): 492–98.

Deleuze, Gilles. 1986. Course Transcriptions [of *Cours sur Michel Foucault (1985–1986)*]. Edited by N. Morar, T. Nail, and D. Smith. https://www.cla.purdue.edu/research/deleuze/Course%20Transcriptions.html.

———. 1988. *Foucault*. Translated by S. Hand. Minneapolis: University of Minnesota Press. Originally published as *Foucault*, 1986.

———. 1995a. "Control and Becoming." In *Negotiations, 1972–1990*, translated by M. Joughin, 169–76. New York: Columbia University Press. Originally published in 1990.

———. 1995b. "Postscript on Control Societies." In *Negotiations*, translated by M. Joughin, 177–82. New York: Columbia University Press. Originally published in 1990.

Deleuze, Gilles, and Félix Guattari. 1994. *What Is Philosophy?* Translated by G. Burchell and H. Tomlinson. New York: Columbia University Press. Originally published as *Qu'est-ce que la philosophie?*, 1991.

DeLillo, Don. 1986. *White Noise*. New York: Penguin Books.

Deutscher, Penelope. 2017. *Foucault's Futures: A Critique of Reproductive Reason*. New York: Columbia University Press.

Dewey, John. (1927) 2016. *The Public and Its Problems*. Edited by M. Rogers. Athens, OH: Swallow Press.

———. (1940) 1963. *The Living Thoughts of Thomas Jefferson Presented by John Dewey*. New York: Premier/Fawcett.

Dick, Philip K. (1955) 1987. "Autofac." In *The Minority Report and Other Classic Stories*, 1–20. New York: Citadel Press.

Dilts, Andrew. 2014. *Punishment and Inclusion: Race, Membership, and the Limits of American Liberalism*. New York: Fordham University Press.

Dreyfus, Hubert L., and Paul Rabinow. 1983. *Michel Foucault: Beyond Structuralism and Hermeneutics*, 2nd ed. Chicago: University of Chicago Press.

Dublin, Louis. 1913. *The Registration of Vital Statistics and Good Business*. New York: Metropolitan Life Insurance Company.

Du Bois, W. E. B. (1903) 1989. *The Souls of Black Folk*. New York: Penguin.

Dumont, Frank. 2010. *A History of Personality Psychology: Theory Science, and Research from Hellenism to the Twenty-First Century*. Cambridge: Cambridge University Press.

Dunn, Halbert L. 1935. "Development of Vital Statistics in the Bureau of the Census." *American Journal of Public Health* 25, no. 12 (December): 1321–26.

———, ed. 1954. "History and Organization of the Vital Statistics System." *Vital Statistics of the United States, 1950, Volume 1*. Washington: Government Printing Office.

Duran, Jane. 2014. "Ellen Gates Star and Julia Lathrop: Hull House and Philosophy." *Pluralist* 9, no. 1 (Spring): 1–13.

Ehler, Elmer W. 1912. "The Legal Importance of Birth Registration." In *Transactions of the Third Annual Meeting, American Association for the Study and Prevention of Infant Mortality*, 102–8. Baltimore: Franklin.

Eliot, T. S. 1934. *The Rock: A Pageant Play*. New York: Harcourt, Brace.

Engerman, Stanley L. 2012. "Monitoring the Abolition of the International Slave Trade: Slave Registration in the British Caribbean." In Breckenridge and Szreter 2012, 323–34.

Erickson, Paul, Judy L. Klein, Lorraine Daston, Rebecca Lemov, Thomas Sturm, and Michael D. Gordin. 2013. *How Reason Almost Lost Its Mind: The Strange Career of Cold War Rationality*. Chicago: University of Chicago Press.

Erlenbusch-Anderson, Verena. 2018. *Genealogies of Terrorism: Revolution, State Terror, Empire*. New York: Columbia University Press.

Ernst, Wolfgang. 2013. *Digital Memory and the Archive*. Minneapolis: University of Minnesota.

Esposito, Roberto. 2008. *Bíos: Biopolitics and Philosophy*. Translated by T. Campbell. Minneapolis: University of Minnesota Press. Originally published as *Bíos: Biopolitica e filosofia*, 2004.

Eysenck, Hans, and Sybil Eysenck. 1976. *Eysenck Personality Questionnaire*. San Diego: Educational and Industrial Testing Service.

Fahmy, Khaled. 2012. "Birth of the 'Secular' Individual: Medical and Legal Methods of Identification in Nineteenth-Century Egypt." In Breckenridge and Szreter 2012, 335–56.

Farge, Arlette. 2013. *The Allure of the Archives*. Translated by T. Scott-Railton. New Haven: Yale University Press. Originally published as *Le goût de l'archive*, 1989.

Farge, Arlette, and Michel Foucault. 2016. *Disorderly Families: Infamous Letters from the Bastille Archives*. Edited by N. Luxon, translated by T. Scott-Railton. Minneapolis: University of Minnesota Press. Originally published as *Le désordre des familles: Lettres de cachet des Archives de la Bastille au XVIIIe siècle*, 1982.

Favinger, Calvin E., and Daniel A. Wilcox. 1939. *Social Security Taxation and Records*. New York: Prentice-Hall.

Fay, Joseph L., and Max J. Wasserman. 1938. "Accounting Operations of the Bureau of Old-Age Insurance." *Social Security Bulletin* 1, no. 6 (June): 24–28.

Fichte, Johann Gottlieb. 2000. *Foundations of Natural Right: According to the Principles of the Wissenschaftslehre*. Edited by F. Neuhouser, translated by M. Baur. Cambridge: Cambridge University Press. Originally published as *Grundlage des Naturrechts nach Principien der Wissenschaftslehre*, 1796–97.

Fisher, Ernest. 1923a. *Principles of Real Estate Practice*. New York: Macmillan.

———. 1923b. "Real Estate Courses for Real Estate Boards and Educational Institutions." *National Real Estate Journal* 24, no. 27 (December 31): 11–17.

———. 1924. "Farm Land Appraisal Practice." In *Annals of Real Estate Practice*, 87–97. Chicago: National Association of Real Estate Boards.

———. 1925. "A Review of Real Estate Literature." In *Annals of Real Estate Practice*, 1: 243–57. Chicago: National Association of Real Estate Boards.

———. 1930. *Advanced Principles of Real Estate Practice*. New York: Macmillan.

Floridi, Luciano. 2003. "Information." In *The Blackwell Guide to the Philosophy of Computing and Information*, edited by L. Floridi, 40–62. Malden: Blackwell.

———. 2010. *Information: A Very Short Introduction*. Oxford: Oxford University Press.

Foucault, Michel. 1971. "Nietzsche, Genealogy, History." In Foucault 1998, 369–91.

———. 1973. *The Birth of the Clinic: An Archaeology of Medical Perception*. Translated by A. M. Sheridan Smith. New York: Vintage. Originally published as *Naissance de la clinique*, 1963.

———. 1977. "Truth and Power." In Foucault 2000, 111–33.

———. 1978. "Questions of Method (Round Table of 20 May 1978)." In Foucault 2000, 223–38.

———. 1980. "The Confession of the Flesh." In *Power/Knowledge: Selected Interviews and Other Writings, 1972–1977*, edited by C. Gordon, 194–228. New York: Pantheon. Originally published in 1977.

———. 1982. "The Subject and Power." In Foucault 2000, 326–48.

———. 1983. "Self Writing." In Foucault 1997, 207–21.

———. 1988. "Power, Moral Values, and the Intellectual." Interview by Michael Bess. *History of the Present*, spring, 1–2, 11–13. https://www.michaelbess.org/foucault-interview/. Originally published in 1980.

———. 1990. *The History of Sexuality, Volume 1: An Introduction*. Translated by R. Hurley. New York: Vintage Books. Originally published as *Histoire de la sexualité I: La volonté de savoir*, 1976.

———. 1994. "Message ou bruit?" In *Dits et Ecrits, tome 1*, 558–61. Paris: Gallimard. Originally published in 1966.

———. 1995. *Discipline and Punish: The Birth of the Prison*. Translated by A. Sheridan. New York: Vintage Books. Originally published as *Surveiller et punir: Naissance de la prison*, 1975.

———. 1997. *Ethics: Subjectivity, and Truth*. Edited by P. Rabinow. Translated by R. Hurley and others. Vol. 1 of Essential Works of Michel Foucault, 1954–1984. New York: New Press.

———. 1998. *Aesthetics, Method, and Epistemology*. Edited by J. Faubion and P. Rabinow. Translated by R. Hurley and others. Vol. 2 of Essential Works of Michel Foucault, 1954–1984. New York: New Press.

———. 2000. *Power*. Edited by J. Faubion and P. Rabinow. Translated by R. Hurley and others. Vol. 3 of Essential Works of Michel Foucault, 1954–1984. New York: New Press.

———. 2003a. *Abnormal: Lectures at the Collège de France, 1974–1975*. Edited by V. Marchetti and A. Salomoni. Translated by G. Burchell. Series edited by A. Davidson. New York: Picador. Originally presented orally as *Les anormaux*, 1975.

———. 2003b. *"Society Must be Defended": Lectures at the Collège de France, 1975–1976.* Edited by M. Bertani and A. Fontana. Translated by D. Macey. Series edited by A. Davidson. New York: Picador. Originally presented orally as *"Il faut défendre la société"*, 1976.

———. 2006. *Psychiatric Power: Lectures at the Collège de France, 1973–1974.* Edited by J. Lagrange. Translated by G. Burchell. Series edited by A. Davidson. New York: Picador. Originally presented orally as *Le pouvoir psychiatrique*, 1974.

———. 2007. *Security, Territory, Population: Lectures at the Collège de France, 1977–1978.* Edited by M. Senellart. Translated by G. Burchell. Series edited by A. Davidson. New York: Palgrave Macmillan. Originally presented orally as *Sécurité, territoire, population*, 1978.

———. 2008. *The Birth of Biopolitics: Lectures at the Collège de France, 1978–79.* Edited by M. Senellart. Translated by G. Burchell. Series edited by A. Davidson. New York: Palgrave Macmillan. Originally presented orally as *Naissance de la biopolitique*, 1979.

———. 2011. *The Government of Self and Others: Lectures at the Collège de France, 1982–83.* Edited by F. Gros. Translated by G. Burchell. Series edited by A. Davidson. New York: Palgrave Macmillan. Originally presented orally as *Le gouvernement de soi et des autres*, 1983.

———. 2012. "The Incorporation of the Hospital into Modern Technology." Translated by E. Knowlton Jr., W. J. King, and S. Elden. In *Space, Knowledge and Power: Foucault and Geography*, edited by J. W. Crampton and S. Elden, 141–51. Abingdon: Ashgate. Originally published in 1974.

———. 2015. *The Punitive Society: Lectures at the Collège de France, 1972–1973.* Edited by B. E. Harcourt. Translated by G. Burchell. Series edited by A. Davidson. New York: Palgrave Macmillan. Originally presented orally as *La société punitive*, 1973.

Foucault, Michel, and Gilles Deleuze. 1980. "Intellectuals and Power." In *Language, Counter-Memory, Practice: Selected Essays and Interviews*, by M. Foucault, edited by D. F. Bouchard, translated by D. F. Bouchard and S. Simon, 205–17. Ithaca: Cornell University Press. Originally published in 1972.

Franz, Shepherd Ivory. 1912. *Handbook of Mental Examination Methods.* New York: Journal of Nervous and Mental Disease Publishing.

———. 1920. *Handbook of Mental Examination Methods.* 2nd ed. New York: Macmillan.

French, Martin A., and Simone Browne. 2014. "Profiles and Profiling Technology: Stereotypes, Surveillance, and Governmentality." In *Criminalization, Representation, Regulation: Thinking Differently about Crime*, edited by D. Brock, A. Glasbeek, and C. Murdocca, 251–80. Toronto: University of Toronto Press.

Freud, Sigmund. 1978. *The Question of Lay Analysis: Conversations with an Impartial Person.* Edited and translated by J. Strachey. New York: W. W. Norton. Originally published as *Die Frage der Laienanalyse*, 1926.

Freund, David M. P. 2007. *Colored Property: State Policy and White Racial Politics in Suburban America.* Chicago: University of Chicago Press.

Funder, David C. 1997. *The Personality Puzzle.* New York: Norton.

Galison, Peter. 1994. "The Ontology of the Enemy: Norbert Wiener and the Cybernetic Vision." *Critical Inquiry* 21, no. 1 (Autumn): 228–66.

Galloway, Alexander. 2004. *Protocol: How Control Exists after Decentralization.* Cambridge: MIT Press.

———. 2012. *The Interface Effect.* London: Polity.

Galloway, Alexander, and Eugene Thacker. 2007. *The Exploit: A Theory of Networks.* Minneapolis: University of Minnesota Press.

Galton, Francis. 1879. "Psychometric Experiments." *Brain* 2:149–62.

———. (1883) 1919. *Inquiries into Human Faculty and Its Development*, 2nd ed. New York: E. P. Dutton.

———. 1884. "Measurement of Character." *Fortnightly Review* 36:179–85.

———. 1888. "Co-relations and Their Measurement, Chiefly from Anthropometric Data." *Proceedings of the Royal Society of London* 45:135–45.

Gans, Herbert. (1967) 2017. *The Levittowners: Ways of Life and Politics in a New Suburban Community*. New York: Columbia University Press.

Gehl, Robert W. 2018. *Weaving Dark Webs: Violence, Propriety, Authenticity*. Cambridge: MIT Press.

Geoghegan, Bernard Dionysus. 2011. "From Information Theory to French Theory: Jakobson, Lévi-Strauss, and the Cybernetic Apparatus." *Critical Inquiry* 38 (Autumn): 96–126.

Gillespie, Tarleton. 2014. "The Relevance of Algorithms." In *Media Technologies: Essays on Communication, Materiality, and Society*, edited by T. Gillespie, P. Boczkowski, and K. Foot, 167–93. Cambridge: MIT Press.

Gitelman, Lisa. 1999. *Scripts, Grooves, and Writing Machines: Representing Technologies in the Edison Era*. Stanford: Stanford University Press.

———. 2014. *Paper Knowledge: Toward a Media History of Documents*. Durham: Duke University Press.

Golumbia, David. 2009. *The Cultural Logic of Computation*. Cambridge: Harvard University Press.

Gould, Stephen Jay. (1981) 1996. *The Mismeasure of Man*. New York: W. W. Norton.

Gray, Charles. 1931. "New Scientific System for Listing Real Estate." *National Real Estate Journal* 32, no. 21 (October 12): 11–17.

Gregg, Melissa. 2015. "The Gift That Is Not Given." In *Data, Now Bigger and Better!*, edited by T. Boellstorff and B. Maurer, 47–66. Chicago: Prickly Paradigm.

Groebner, Valentin. 2007. *Who Are You? Identification, Deception, and Surveillance in Early Modern Europe*. Translated by M. Kyburz and J. Peck. New York: Zone Books.

Gutman, Robert. 1958. "Birth and Death Registration in Massachusetts," part 1 of 4. *Milbank Memorial Fund Quarterly* 36, no. 1 (January): 58–74.

Habermas, Jürgen. 1985. *The Theory of Communicative Action*. 2 vols. Translated by T. McCarthy. Boston: Beacon. Originally published as *Theorie des kommunikativen Handelns*, 1981.

———. 1990. *Moral Consciousness and Communicative Action*, translated by C. Lenhardt and S. Nicholsen. Cambridge: MIT Press. Originally published as *Moralbewusstsein und kommunikatives Handeln*, 1983.

———. 1991. *The Structural Transformation of the Public Sphere*. Translated by T. Burger with F. Lawrence. Cambridge: MIT Press. Originally published as *Strukturwandel der Öffentlichkeit*, 1962.

———. 1996. *Between Facts and Norms: Contributions to a Discourse Theory of Law and Democracy*. Translated by W. Rehg. Cambridge: MIT Press. Originally published as *Faktizität und Geltung*, 1992.

Hacking, Ian. 1986. "Making Up People." In *Reconstructing Individualism: Autonomy, Individuality, and the Self in Western Thought*, edited by T. Heller, M. Sosna, and D. Wellbery, 222–36. Stanford: Stanford University Press.

———. (1990) 2006. *The Taming of Chance*. Cambridge: Cambridge University Press.

———. 1995. *Rewriting the Soul: Multiple Personality and the Sciences of Memory*. Princeton: Princeton University Press.

———. 2002. "Michel Foucault's Immature Science." In *Historical Ontology*, 87–98. Cambridge: Harvard University Press, 2002. Originally published in 1979.

———. 2009. *Scientific Reason*. Edited by D. Yeh and J. Yuann. Taipei: National Taiwan University Press.

———. 2015. "Biopower and the Avalance of Printed Numbers." In *Biopower: Foucault and Beyond*, edited by V. Cisney and N. Morar, 65–81. Chicago: University of Chicago Press. Originally published in 1982.

Haggerty, Kevin, and Richard Ericson. 2000. "The Surveillant Assemblage." *British Journal of Sociology* 51, no. 4: 605–22.

Halpern, Orit. 2014. *Beautiful Data: A History of Vision and Reason since 1945*. Durham: Duke University Press.

Haltunnen, Karen. 1989. "From Parlor to Living Room: Domestic Space, Interior Decoration, and the Culture of Personality." In *Consuming Visions: Accumulation and the Display of Goods in America*, edited by S. Bronner, 157–90. New York: W. W. Norton.

Hansen, Mark B. N. 2015. "Symbolizing Time: Kittler and Twenty-First Century Media." In Sale and Salisbury 2015, 210–37.

Haraway, Donna. 1991. "A Cyborg Manifesto: Science, Technology, and Socialist-Feminism in the Late Twentieth Century." In *Simians, Cyborgs, and Women: The Reinvention of Nature*, 149–81. New York: Routledge. Originally published in 1985.

Harcourt, Bernard E. 2007. *Against Prediction: Profiling, Policing, and Punishing in an Actuarial Age*. Chicago: University of Chicago Press, 2007.

———. 2013. "Rethinking Power with and beyond Foucault." in *Carceral Notebooks* 9:79–87.

———. 2015. *Exposed: Desire and Disobedience in the Digital Age*. Cambridge: Harvard University Press.

Harriss, C. Lowell. 1951. *History and Policies of the Home Owners' Loan Corporation*. New York: National Bureau of Economic Research.

Hawkins, W. Ashbie. 1911. "A Year of Segregation in Baltimore" in *Crisis* 3 (November): 28.

Hayles, N. Katherine. 1999. *How We Became Posthuman: Virtual Bodies in Cybernetics, Literature, and Informatics*. Chicago: University of Chicago Press.

———. 2005. *My Mother Was a Computer: Digital Subjects and Literary Texts*. Chicago: University of Chicago Press.

———. 2010. "Cybernetics." In Mitchell and Hansen 2010, 145–56.

Headrick, Daniel R. 2002. *When Information Came of Age: Technologies of Knowledge in the Age of Reason and Revolution, 1700–1850*. Oxford: Oxford University Press.

Heidegger, Martin. 1977. "The End of Philosophy and the Task of Thinking." In *Basic Writings: From Being and Time (1927) to The Task of Thinking (1964)*, edited by D. Krell, 373–92. San Francisco: HarperCollins. Originally published in 1966.

Heims, Steve J. 1993. *Constructing a Social Science for Postwar America: The Cybernetics Group, 1946–1953*. Cambridge: MIT Press.

Helper, Rose. 1969. *Racial Policies and Practices of Real Estate Brokers*. Minneapolis: University of Minnesota Press.

Hemenway, Henry B., William H. Davis, and Charles V. Chapin. 1928. "Definition of Stillbirth." *American Journal of Public Health and the Nation's Health* 18, no. 1 (January): 25–32.

Herman, Ellen. 1995. *The Romance of American Psychology: Political Culture in the Age of Experts*. Berkeley: University of California Press.

Hervey, Ginger. 2017. "Justice Evades Slovenia's 'Erased' Citizens." *Politico*, March 28. http://www.politico.eu/article/justice-evades-slovenia-erased-citizens-yugoslavia/.

Hetzel, Alice M. 1997. *U.S. Vital Statistics System: Major Activities and Developments, 1950–95*. Hyattsville, MD: National Center for Health Statistics.

Higgs, Edward. 2004. *The Information State in England: The Central Collection of Information on Citizens, 1500–2000*. London: Palgrave.

———. 2011. *Identifying the English: A History of Personal Identification 1500 to the Present*. London: Continuum.

Hillier, Amy E. 2003. "Redlining and the Homeowners' Loan Corporation." *Journal of Urban History* 29, no. 4: 394–420.

Hollingworth, H. L. 1920. *The Psychology of Functional Neuroses*. New York: D. Appleton.

Home Owners' Loan Corporation. 1933. *Loan Regulations* (Form 7). Washington: Government Printing Office.

Horkheimer, Max, and Theodor Adorno. (1944) 2002. *The Dialectic of Enlightenment*. Stanford: Stanford University Press.

Howe, Frederic. 1911. "The Scientific Appraisal of Real Estate as a Basis for Taxation." *City Club Bulletin of Philadelphia* 3–4:98–119.

Hu, Tung-Hui. 2015. *A Prehistory of the Cloud*. Cambridge: MIT Press.

Huhtamo, Erkki. 2013. *Illusions in Motion: Media Archaeology of the Moving Panorama and Related Spectacles*. Cambridge: MIT Press.

Hull, Gordon. 2013. "Biopolitics Is Not (Primarily) about Life: On Biopolitics, Neoliberalism, and Families." *Journal of Speculative Philosophy* 27, no. 3 (Fall): 322–35.

Hurd, Richard M. 1903. *Principles of City Land Value*. New York: Record and Guide.

Hurty, J. N. 1910a. "The Bookkeeping of Humanity." *Michigan Monthly Bulletin of Vital Statistics* 13, no. 10 (October): 75–80.

———. 1910b. "The Bookkeeping of Humanity." *Journal of the American Medical Association* 55, no. 14 (October): 1157–60.

Igo, Sarah E. 2007. *The Averaged American: Surveys, Citizens, and the Making of a Mass Public*. Cambridge: Harvard University Press.

Jabko, Nicolas, and Adam Sheingate. 2018. "Practices of Dynamic Order." *Perspectives on Politics* 16, no. 2 (June): 312–27.

Jackson, Kenneth T. 1980. "Race, Ethnicity, and Real Estate Appraisal: The Home Owners Loan Corporation and the Federal Housing Administration." *Journal of Urban History* 6, no. 4 (August): 419–52.

———. 1985. *Crabgrass Frontier: The Suburbanization of the United States*. New York: Oxford University Press.

James, William. (1890) 1950. *The Principles of Psychology*. 2 vols. New York: Dover Press, 1950.

———. 1984. "Person and Personality: From *Johnson's Universal Cyclopaedia*." In *Essays in Psychology*, 315–22. Cambridge: Harvard University Press. Originally published in 1895.

Jansen, Sue Curry. 2009. "Phantom Conflict: Lippmann, Dewey, and the Fate of the Public in Modern Society." *Communication and Critical/Cultural Studies* 6, no. 3: 221–45.

Johnson, James Weldon. (1912) 1995. *The Autobiography of an Ex-colored Man*. New York: Dover.

Kafka, Ben. 2012. *The Demon of Writing: Powers and Failures of Paperwork*. Cambridge: MIT Press.

Kaplan, Jonathan, and Andrew Valls. 2007. "Housing Discrimination as a Basis for Black Reparations." *Public Affairs Quarterly* 21, no. 3: 255–73.

Kay, Lily. 2000. *Who Wrote the Book of Life? A History of the Genetic Code*. Stanford: Stanford University Press.

Keller, Evelyn Fox. 1995. *Refiguring Life: Metaphors of Twentieth-Century Biology*. New York: Columbia University Press.

Kirschenbaum, Matthew G. 2008. *Mechanisms: New Media and the Forensic Imagination*. Cambridge: MIT Press.

Kitchin, Rob. 2014. *The Data Revolution: Big Data, Open Data, Data Infrastructures and Their Consequences*. London: Sage.

Kittler, Friedrich. 1982. "The God of Ears." Translated by P. Fiegelfeld and A. Moore. In Sale and Salisbury 2015, 3–21.

———. 1990. *Discourse Networks, 1800/1900*. Translated by M. Metteer with C. Cullens. Stanford: Stanford University Press. Originally published as *Aufschreibesysteme 1800/1900*, 1985.

———. 1999. *Gramophone, Film, Typewriter*. Translated by G. Winthrop-Young and M. Wutz. Stanford: Stanford University Press. Originally published as *Grammophon Film Typewriter*, 1986.

———. 2006. "Cold War Networks or Kaiserstr. 2, Neubabelsberg." In *New Media, Old Media: A History and Theory Reader*, edited by W. Chun and T. Keenan, translated by T. Krapp, 181–86. New York: Routledge. Originally delivered in 1996.

———. 2010. *Optical Media: Berlin Lectures 1999*. Translated by A. Enns. Cambridge: Polity. Originally published as *Optische Medien*, 2002.

———. 2013. "The Artificial Intelligence of World War: Alan Turing." In *The Truth of the Technological World: Essays on the Genealogy of Presence*, translated by E. Butler, 178–94. Stanford: Stanford University Press. Originally published in 1990.

Kline, Ronald R. 2015. *The Cybernetics Moment: Or Why We Call Our Age the Information Age*. Baltimore: Johns Hopkins University Press.

Koopman, Colin. 2009. *Pragmatism as Transition: Historicity and Hope in James, Dewey, and Rorty*. New York: Columbia University Press.

———. 2013. *Genealogy as Critique: Foucault and the Problems of Modernity*. Bloomington: Indiana University Press.

———. 2014. "Michel Foucault's Critical Empiricism Today: Concepts and Analytics in the Critique of Biopower and Infopower." In *Foucault Now: Current Perspectives in Foucault Studies*, edited by J. Faubion, 88–111. London: Polity Press.

———. 2015a. "The Algorithm and the Watchtower." *New Inquiry*, September 29. https://thenewinquiry.com/the-algorithm-and-the-watchtower/.

———. 2015b. "Two Uses of Michel Foucault in Political Theory: Concepts and Methods in Giorgio Agamben and Ian Hacking." *Constellations* 22, no. 4 (December): 571–85.

———. 2016. "Transforming the Self amidst the Challenges of Chance: William James on 'Our Undisciplinables.'" *Diacritics* 44, no. 4: 40–65.

———. 2017a. "Conceptual Analysis for Genealogical Philosophy: How to Study the History of Practices after Foucault and Wittgenstein." *Southern Journal of Philosophy* 55, Special Supp. on "Critical Histories of the Present," edited by V. Erlenbusch (September): 103–21.

———. 2017b. "The Power Thinker." *Aeon*, March 15. https://aeon.co/essays/why-foucaults-work-on-power-is-more-important-than-ever.

———. 2018. "How Democracy Can Survive Big Data." *New York Times*, March 22. https://www.nytimes.com/2018/03/22/opinion/democracy-survive-data.html.

———. Forthcoming. "Coding the Self: The Infopolitics and Biopolitics of Genetic Sciences." *Hastings Center Report.*

Koopman, Colin, and Tomas Matza. 2013. "Putting Foucault to Work: Analytic and Concept in Foucaultian Inquiry." *Critical Inquiry* 39, no. 4 (Summer): 817–40.

Krafft-Ebing, Richard von. 1998. *Psycopathia Sexualis.* Translated by F. Klaf. New York: Arcade. Originally published as *Psycopathia Sexualis,* 1886.

Krajewski, Markus. 2011. *Paper Machines: About Cards and Catalogs, 1548–1929.* Translated by P. Krapp. Cambridge: MIT Press. Originally published as *Zettelwirtschaft: die Geburt der Kartei aus dem Geiste der Bibliothek,* 2002.

Krämer, Sybille. 2006. "Time Axis Manipulation: On Friedrich Kittler's Conception of Media." *Theory, Culture, and Society* 23, no. 7–8: 93–109.

Krause, Sharon R. 2008. *Civil Passions: Moral Sentiment and Democratic Deliberation.* Princeton: Princeton University Press.

Laird, Donald A. 1925. "A Mental Hygiene and Vocational Test." *Journal of Educational Psychology* 16, no. 6 (September): 419–22.

Landrum, Shane. 2010. "Undocumented Citizens: The Crisis of U.S. Birth Certificates, 1940–1945." Paper presented at the American Historical Association Annual Meeting, January 8. http://cliotropic.org/blog/talks/undocumented-citizens-aha-2010/.

———. 2015. "From Family Bibles to Birth Certificates: Young People, Proof of Age, and American Political Cultures, 1820–1915." In *Age in America,* edited by C. Field and N. Syrett, 124–27. New York: New York University Press.

Lange, Alethea, and Rena Coen. 2016. "How Does the Internet Know Your Race?" Blog post, Center for Democracy and Technology, September 7. https://cdt.org/blog/how-does-the-internet-know-your-race/.

Larsen, Randy, and David Buss. 2010. *Personality Psychology.* 4th ed. New York: McGraw-Hill.

Lathrop, Julia. (1895) 2007. "The Cook County Charities." In *Hull-House Maps and Papers,* by the residents of Hull-House, 120–29. Urbana: University of Illinois Press.

———. 1914a. *First Annual Report of the Children's Bureau to the Secretary of Labor for the Fiscal Year Ended June 30, 1913.* Washington: Government Printing Office, 1914.

———. 1914b. *Second Annual Report of the Children's Bureau to the Secretary of Labor, Fiscal Year Ended June 30, 1914.* Washington: Government Printing Office.

———. 1915. *Third Annual Report of the Chief, Children's Bureau to the Secretary of Labor, Fiscal Year Ended June 30, 1915.* Washington: Government Printing Office.

———. 1916. *Fourth Annual Report of the Children's Bureau to the Secretary of Labor, Fiscal Year Ended June 30, 1916.* Washington: Government Printing Office.

———. 1917. *Fifth Annual Report of the Children's Bureau to the Secretary of Labor, Fiscal Year Ended June 30, 1917.* Washington: Government Printing Office.

Latour, Bruno. 1986. "The Powers of Association." In *Power, Action, and Belief,* edited by John Law, 264–80. London: Routledge & Kegan Paul.

———. 1987. *Science in Action: How to Follow Scientists and Engineers through Society.* Cambridge: Harvard University Press.

———. 1999. *Pandora's Hope: Essays on the Reality of Science Studies.* Cambridge: Harvard University Press.

———. 2004. "Why Has Critique Run out of Steam? From Matters of Fact to Matters of Concern." *Critical Inquiry* 30, no. 2 (Winter): 225–48.

Lazzarato, Maurizio. 2006. "The Concepts of Life and the Living in the Societies of Control." In

Deleuze and the Social, edited by M. Fuglsang and B. Sørensen, 171–90. Edinburgh: Edinburgh University Press.

Leary, David E. 1990. "William James on the Self and Personality." In *Reflections on "The Principles of Psychology": William James after a Century*, edited by M. G. Johnson and T. B. Henley, 101–37. Hillsdale: Lawrence Erlbaum.

Lemov, Rebecca. 2005. *World as Laboratory: Experiments with Mice, Mazes, and Men.* New York: Hill and Wang.

———. 2009. "Towards a Data Base of Dreams: Assembling an Archive of Elusive Materials, c. 1947–61." *History Workshop Journal* 67:44–68.

Lenhart, Robert F. 1943. "Completeness of Birth Registration in the United States in 1940." *American Journal of Public Health and the Nation's Health* 33, no. 6 (June): 685–90.

Leonelli, Sabina. 2016. *Data-centric Biology: A Philosophical Study.* Chicago: University of Chicago Press.

Lerrigo, Charles H. 1922. "Simplicity in Preparation of Blanks and Forms." *American Journal of Public Health* 12, no. 2 (February): 116–19.

Lessig, Lawrence. 2004. *Free Culture: How Big Media Uses Technology and the Law to Lock Down Culture and Control Creativity.* New York: Penguin. Available online at http://www.free-culture.cc/freeculture.pdf.

———. 2006. *Code, Version 2.0.* New York: Basic Books. Available online at http://codev2.cc/download+remix/Lessig-Codev2.pdf.

Levy, Mark. 1931. "Small Store Apartment and Office Property in Outlying District of Shifting Values." in *National Real Estate Journal* 32, no. 2 (January 19): 12–16.

Lewis, Sinclair. (1922) 1961. *Babbitt.* New York: Signet Classics.

Lindenmeyer, Kriste. 1997. *"A Right to Childhood": The U.S. Children's Bureau and Child Welfare, 1912–46.* Urbana: University of Illinois Press.

Lippmann, Walter. (1913) 1969. *A Preface to Politics.* Ann Arbor: University of Michigan Press.

———. (1914) 1985. *Drift and Mastery.* Madison: University of Wisconsin Press.

———. 1922. "The Abuse of the Tests." Part 4 of 6. *New Republic*, November 15, 297–98.

———. (1922) 1997. *Public Opinion.* New York: Free Press.

———. (1925) 1993. *The Phantom Public.* New Brunswick: Transaction Publishers.

———. 1955. *The Public Philosophy.* Boston: Little, Brown.

Littlefield, Daniel F. Jr., and Lonnie E. Underhill. 1971. "Renaming the American Indian: 1890–1913." *American Studies* 12, no. 2 (Fall): 33–45.

Livingston, James. 2001. *Pragmatism, Feminism, and Democracy: Rethinking the Politics of American History.* New York: Routledge.

Lorenzini, Daniele. 2016. "Foucault, Regimes of Truth and the Making of the Subject." In *Foucault and the Making of Subjects*, edited by L. Cremonesi, O. Irrera, D. Lorenzini, and M. Tazzioli, 63–76. New York: Rowman & Littlefield International.

Lyon, David. 2009. *Identifying Citizens: ID Cards as Surveillance.* New York: Polity.

———. 2010. "Surveillance, Power and Everyday Life." In *Emerging Digital Spaces in Contemporary Society*, edited by P. Kalantzis-Cope and K. Gherab-Martín, 107–20. London: Palgrave Macmillan.

MacChesney, Nathan William. 1924a. *Brief and Argument for Protection of Term "Realtor."* Chicago: National Association of Real Estate Boards.

———. 1924b. *Pleadings: Suggested Form of Pleadings in Suit to Protect the Use of the Term "Realtor" by Local Real Estate Board and the National Association of Real Estate Boards.* Chicago: National Association of Real Estate Boards.

———. 1927. *The Principles of Real Estate Law*. New York: Macmillan.

Mader, Mary Beth. 2007. "Foucault and Social Measure." *Journal of French Philosophy* 17, no. 1 (Spring): 1–25.

———. 2011. *Sleights of Reason: Norm, Bisexuality, Development*. Albany: State University of New York Press.

Manson, Grace E. 1926. *A Bibliography of the Analysis and Measurement of Human Personality up to 1926*. Washington, DC: National Research Council.

Marres, Noortje. 2012. *Material Participation*. New York: Palgrave Macmillan.

Marshall, Dominique. 2012. "Birth Registration and the Promotion of Children's Rights in the Interwar Years: The Save the Children's International Union's Conference on the African Child, and Herbert Hoover's American Child Health Association." In Breckenridge and Szreter 2012, 449–74.

Massey, Douglas, and Nancy Denton. 1993. *American Apartheid: Segregation and the Making of the Underclass*. Cambridge: Harvard University Press.

Matthews, Ellen. 1923. "A Study of Emotional Stability in Children: By Means of a Questionnaire." *Journal of Delinquency* 8, no. 1 (January): 1–40.

May, Mark A., and Hugh Hartshorne. 1926. "Personality and Character Tests." *Psychological Bulletin* 23, no. 7: 395–411.

McCarthy, Thomas. 2004. "Coming to Terms with Our Past, Part II: On the Morality and Politics of Reparations for Slavery." *Political Theory* 32, no. 6 (December): 750–72.

McKeon, Richard. 1990. "Communication, Truth, and Society." In *Freedom and History and Other Essays*, edited by Z. McKeon, 88–102. Chicago: University of Chicago Press. Originally published in 1957.

McKinley, Charles, and Robert W. Frase. 1970. *Launching Social Security: A Capture-and-Record Account, 1935–1937*. Madison: University of Wisconsin Press.

McMichael, Stanley. 1931. *McMichael's Appraising Manual: A Real Estate Appraising Handbook for Field Work and Advanced Study Courses*, 1st ed. New York: Prentice-Hall.

———. 1937. *McMichael's Appraising Manual: A Real Estate Appraising Handbook for Field Work and Advanced Study Courses*, 2nd ed. New York: Prentice-Hall.

McMichael, Stanley, and Robert F. Bingham. 1923. *City Growth and Values*. Cleveland: Stanley McMichael Publishing.

McPherson, James Alan. 1972. "'In My Father's House There Are Many Mansions—And I'm Going to Get Me Some of Them Too': The Story of the Contract Buyers League." *Atlantic*, April, 51–82.

McWhorter, Ladelle. 2009. *Racism and Sexual Oppression in Anglo-America: A Genealogy*. Bloomington: Indiana University Press.

———. 2017. "The Morality of Corporate Persons." In "Critical Histories of the Present," edited by V. Erlenbusch. Special supplement, *Southern Journal in Philosophy* 55 (September): 126–48.

Mead, George Herbert. 1938. *The Philosophy of the Act*. Edited by C. Morris. Chicago: University of Chicago Press.

Medina, Eden. 2011. *Cybernetic Revolutionaries: Technology and Politics in Allende's Chile*. Cambridge: MIT Press.

Menninger, Karl. 1930. *The Human Mind*. New York: Alfred Knopf.

Mertzke, Arthur. 1927. *Real Estate Appraising*. Chicago: National Association of Real Estate Boards.

Mindell, David A. 2002. *Between Human and Machine: Feedback, Control, and Computing before Cybernetics*. Baltimore: Johns Hopkins University Press.

Misak, Cheryl. 2013. *The American Pragmatists*. Oxford: Oxford University Press.

———. 2017. "A Pragmatist Account of Legitimacy and Authority: Holmes, Ramsey, and the Moral Force of Law." In *Pragmatism and Justice*, edited by S. Dieleman, D. Rondel, and C. Voparil, 295–308. Oxford: Oxford University Press.

Mischel, Walter. 1968. *Personality and Assessment*. New York: Wiley.

Mitchell, W. J. T., and Mark B. N. Hansen. 2010. *Critical Terms for Media Studies*. Chicago: University of Chicago Press.

Mittelstadt, Brent Daniel, Patrick Allo, Mariarosaria Taddeo, Sandra Wachter, and Luciano Floridi. 2016. "The Ethics of Algorithms: Mapping the Debate." *Big Data and Society* 3, no. 2, (July–December): 1–21.

Monea, Alexander. 2016. "Numerical Mediation and American Governmentality." Ph.D. diss., North Carolina State University.

Monea, Alexander and Packer, Jeremy. 2016. "Media Genealogy and the Politics of Archaeology." *International Journal of Communication*, 10:3141–59.

Moore, Tyler. 2017. "On the Harms Arising from the Equifax Data Breach of 2017." *International Journal of Critical Infrastructure Protection* 19 (December 2017): 47–48.

Mowery, David C., and Timothy Simcoe. 2002. "Is the Internet a US Invention?—an Economic and Technological History of Computer Networking." *Research Policy* 31, nos. 8–9 (December): 1369–87.

Muhammad, Khalil Gibran. 2010. *The Condemnation of Blackness: Race, Crime, and the Making of Modern Urban America*. Cambridge: Harvard University Press.

Nail, Thomas. 2016. "Biopower and Control." In *Between Deleuze and Foucault*, edited by N. Morar, T. Nail, and D. Smith, 247–63. Edinburgh: Edinburgh University Press.

Nakamura, Lisa. 2002. *Cybertypes: Race, Ethnicity, and Identity on the Internet*. New York: Routledge.

National Association of Real Estate Boards. 1924. "Code of Ethics." Rev. ed. *National Real Estate Journal* 25, no. 12 (June 16): 63–64.

National Association of Real Estate Exchanges. 1913. "Ethics of the Real Estate Profession." N.p.: National Association of Real Estate Exchanges. Accessed October 30, 2016. http://www.realtor.org/about-nar/mission-vision-and-history/1913-code-of-ethics.

Nelson, Herbert. 1923. "Systematizing Real Estate Education." *National Real Estate Journal* 24, no. 17 (August 13): 38.

———. 1925. "The Objectives and Content of Real Estate Courses." In *Annals of Real Estate Practice*, 1:195–211. Chicago: National Association of Real Estate Boards.

Nelson, Robert K., LaDale Winling, Richard Marciano, and Nathan Connolly. n.d. "Mapping Inequality: Redlining in New Deal America," *American Panorama*, edited by R. Nelson and E. Ayers. Accessed May 11, 2018. https://dsl.richmond.edu/panorama/redlining/.

Nelson, Thomas. 1925. "A Standard Test in Real Estate." In *Annals of Real Estate Practice*, 1:259–69. Chicago: National Association of Real Estate Boards.

Neurath, Otto. 1936. *International Picture Language: The First Rules of ISOTYPE*. London: Kegan Paul.

———. (1937) 1973. "A New Language." [Original title: "Visual Education."] In *Empiricism and Sociology*, edited by M. Neurath and R. Cohen, 224–26. Boston: D. Reidel.

———. 2010. *From Hieroglyphics to Isotype: A Visual Autobiography*. Edited by M. Eve and C. Burke. London: Hyphen Press.

Newman, Lily Hay. 2017. "Replacing Social Security Numbers Won't Be Easy, but It's Worth It." *Wired*, October 13. https://www.wired.com/story/social-security-number-replacement.

Nicholson, Ian A. M. 2003. *Inventing Personality: Gordon Allport and the Sciences of Selfhood*. Washington: American Psychological Association.

Nicolas, Serge, Bernard Andrieu, Jean-Claude Croizet, Rasyid B. Sanitioso, and Jeremy Trevelyan Burman. 2013. "Sick? Or Slow? On the Origins of Intelligence as a Psychological Object." *Intelligence* 41: 699–711.

Noble, Safiya. 2018. *Algorithms of Oppression: How Search Engines Reinforce Racism*. New York: New York University Press.

Noiriel, Gérard. 2001. "The Identification of the Citizen: The Birth of Republican Civil Status in France." In Caplan and Torpey 2001, 28–48.

Nolan, Preston M. 1925a. "Building Depreciation and Obsolescence." *National Real Estate Journal* 26, no. 15 (July 17): 39–40.

———. 1925b. "Nolanisms for the Appraiser." *National Real Estate Journal* 26, no. 10 (May 18): 34.

Nunberg, Geoff. 2012. "The Informations." Presentation at iConference, University of Toronto, February 9. http://people.ischool.berkeley.edu/%7Enunberg/IConfTalk2-8.pdf.

Olson, Kevin. 2016. *Imagined Sovereignties: The Power of the People and Other Myths of the Modern Age*. Cambridge: Cambridge University Press.

Omi, Michael, and Howard Winant. 1986. *Racial Formation in the United States: From the 1960s to the 1980s*. New York: Routledge.

Orwell, George. 1949. *Nineteen Eighty-Four*. London: Secker and Warburg.

Palmer, George Truman. 1933. "Measurement in Public Health." *Child Health Bulletin* 9, no. 4 (July): 132–34.

Panagia, Davide. 2019. "On the Political Ontology of the *Dispositif*." *Critical Inquiry* 45, no. 3 (forthcoming).

———. n.d. "#Datapolitik: An Introduction." Unpublished manuscript.

Papurt, Maxwell Jerome. 1930. "A Study of the Woodworth Psychoneurotic Inventory with Suggested Revision." *Journal of Abnormal and Social Psychology* 25, no. 3 (October): 335–52.

Parikka, Jussi. 2012. *What is Media Archaeology?* London: Polity Press.

Parker, Jacqueline K., and Edward M. Carpenter. 1981. "Julia Lathrop and the Children's Bureau: The Emergence of an Institution." *Social Service Review* 55, no. 1 (March): 60–77.

Pearson, Susan J. 2015. "'Age Ought to Be a Fact': The Campaign against Child Labor and the Rise of the Birth Certificate." *Journal of American History* 101, no. 4 (March): 1144–65.

Peters, John Durham. 1988. "Information: Notes Toward a Critical History." *Journal of Communication Inquiry* 12, no. 2: 9–23.

———. 1989. "Democracy and American Mass Communication Theory: Dewey, Lippmann, Lazersfeld." *Communication* 11, no. 3: 199–220.

———. 1999. *Speaking into the Air: A History of the Idea of Communication*. Chicago: University of Chicago Press.

———. 2015. *The Marvelous Clouds: Toward a Philosophy of Elemental Media*. Chicago: University of Chicago Press.

Peters, John Durham, and Benjamin Peters. 2016. "Norbert Wiener as Pragmatist." *Empedocles: European Journal for the Philosophy of Communication* 7, no. 2 (October): 157–72.

Philpott, Thomas Lee. 1978. *The Slum and the Ghetto: Neighborhood Deterioration and Middle-Class Reform, Chicago, 1880–1930*. New York: Oxford University Press.

Pickering, Andrew. 2010. *The Cybernetic Brain: Sketches of Another Future*. Chicago: University of Chicago Press.

Pierce, John. (1961) 1980. *An Introduction to Information Theory.* New York: Dover.

Plecker, W. A. 1915. "A Standard Certificate of Birth." *American Journal of Public Health* 5, no. 10 (October): 1044–47.

Poitras, Laura, dir. 2015. *Citizenfour.* Beverly Hills: Anchor Bay Entertainment.

Pollock, W. W., and Karl W. H. Scholz. 1926. *Science and Practice of Urban Land Valuation.* Philadelphia: Manufacturers' Appraisal Company.

———. 1930. *The Personality of a House: The Blue Book of Home Design and Decoration.* New York: Funk & Wagnalls.

Prince, Morton. 1885. *The Nature of Mind and Human Automatism.* Philadelphia: J. B. Lipincott.

———. 1906. *The Dissociation of a Personality: A Biographical Study in Abnormal Psychology.* New York: Longmans, Green.

———. 1914. *The Unconscious: The Fundamentals of Human Personality, Normal and Abnormal.* New York: Macmillan.

———. 1921. *A Critique of Psychanalysis* [*sic*]. Chicago: American Medical Association. Reprinted from *Archives of Neurology and Psychology,* December 6, 610–33.

Prouty, W. L., Clem W. Collins, and Frank H. Prouty. 1930. *Appraisers and Assessors Manual.* New York: McGraw-Hill.

Puckett, Carolyn. 2009. "The Story of the Social Security Number." *Social Security Bulletin* 69, no. 2: 55–74.

Purdon, James. 2016. *Modernist Informatics: Literature, Information, and the State.* Oxford: Oxford University Press.

Rabinow, Paul. 2008. *Marking Time: On the Anthropology of the Contemporary.* Princeton: Princeton University Press.

———. 2011. *The Accompaniment: Assembling the Contemporary.* Chicago: University of Chicago Press.

Rabinow, Paul, and Anthony Stavrianakis. 2014. *Designs on the Contemporary: Anthropological Tests.* Chicago: University of Chicago Press.

Radiolab. 2016. "The Girl Who Doesn't Exist." Produced by WNYC Studios. *Radiolab,* August 29. Podcast, 35:14. http://www.radiolab.org/story/invisible-girl/.

Raley, Rita. 2009. *Tactical Media.* Minneapolis: University of Minnesota Press.

Ramati, Ido, and Amit Pinchevski. 2017. "Uniform Multilingualism: A Media Genealogy of Google Translate." *New Media and Society* 20, no. 7: 2550–65.

Rawls, John. 1971. *A Theory of Justice.* Cambridge: Harvard University Press.

Reeder, F. H. 1937. "What Is a Delayed Certificate and under What Conditions and Requirements Should It Be Filed?" *American Journal of Public Health and the Nation's Health* 27, no. 12 (December): 1216–20.

Reeves, Cuthbert. 1928. *The Appraisal of Urban Land and Buildings: A Working Manual for City Assessors.* New York: Municipal Administration Service.

Reza, Fazlollah. 1961. *An Introduction to Information Theory.* New York: McGraw-Hill.

Rice, Roger L. 1968. "Residential Segregation by Law, 1910–1917." *Journal of Southern History* 34, no. 2 (May): 179–99.

Roback, Abraham A. 1927. *A Bibliography of Character and Personality.* Cambridge: Sci-Art Publishers.

———. 1932–33. "Personality Tests—Whither?" *Character & Personality* 1:214–24.

Robertson, Craig. 2010. *The Passport in America: The History of a Document.* Oxford: Oxford University Press.

Roediger, David. 2006. *Working Toward Whiteness: How America's Immigrants Became White.* New York: Basic Books.

Rogers, Melvin. 2009. *The Undiscovered Dewey.* New York: Columbia University Press.

Rondel, David. 2018. *Pragmatist Egalitarianism.* Oxford: Oxford University Press.

Rorty, Richard. 1989. *Contingency, Irony, and Solidarity.* Cambridge: Cambridge University Press.

Rose, Nikolas. 1985. *The Psychological Complex: Psychology, Politics and Society in England, 1869–1939.* London: Routledge & Kegan Paul.

———. 1996. *Inventing Our Selves: Psychology, Power, and Personhood.* Cambridge: Cambridge University Press.

Rosen, Lawrence. 1995. "The Creation of the Uniform Crime Report: The Role of Social Science." *Social Science History* 19, no. 2 (Summer): 215–38.

Rosenberg, Daniel. 2013. "Data before the Fact." In *"Raw Data" Is an Oxymoron,* edited by L. Gitelman, 15–40. Cambridge: MIT Press.

Rothstein, Richard. 2017. *The Color of Law: A Forgotten History of How Our Government Segregated America.* New York: Liveright.

Rugh, Jacob S. and Douglas S. Massey. 2010. "Racial Segregation and the American Foreclosure Crisis." *American Sociological Review* 75, no. 5 (October): 629–51.

Ruland, Irving. 1920. "The Standardization of Real Estate Valuations." In *Practical Real Estate Methods,* edited by F. Ward, 119–21. New York: Doubleday, Page.

Sale, Stephen, and Salisbury, Laura, eds. 2015. *Kittler Now: Current Perspectives in Kittler Studies.* Cambridge: Polity.

Sandvig, Christian, Kevin Hamilton, Karrie Karahalios, and Cedric Langbort. 2014. "An Algorithm Audit." In *Data and Discrimination: Collected Essays,* edited by S. Gangadharan, 6–10. Washington: New America Foundation.

Satter, Beryl. 2009. *Family Properties: Race, Real Estate, and the Exploitation of Black Urban America.* New York: Picador.

Schlesinger, Arthur M. Jr. 1958. *The Coming of the New Deal: The Age of Roosevelt,* vol. 2. Boston: Houghton Mifflin.

Schudson, Michael. 2008. "The 'Lippmann-Dewey Debate' and the Invention of Walter Lippmann as an Anti-Democrat, 1986–1996." *International Journal of Communication,* 2:1–20.

Schüll, Natasha Dow. 2016. "Data for Life: Wearable Technology and the Design of Self-Care." *BioSocieties* 11, no. 3: 317–33.

———. 2018. "Self in the Loop: Bits, Patterns, and Pathways in the Quantified Self." In *A Networked Self and Human Augmentics, Artificial Intelligence, Sentience,* edited by Z. Papacharissi, 25–38. New York: Routledge.

Scott, James C. 1998. *Seeing Like a State: How Certain Schemes to Improve the Human Condition Have Failed.* New Haven: Yale University Press.

Scott, James C., John Terhanian, and Jeremy Mathias. 2002. "The Production of Legal Identities Proper to States: The Case of the Permanent Family Surname." *Comparative Studies in Society and History* 44, no. 1 (January): 4–44.

Sellars, Wilfrid. (1956) 1997. *Empiricism and the Philosophy of Mind.* Cambridge: Harvard University Press.

Sellin, Thorsten. 1928. "The Negro Criminal: A Statistical Note." *Annals of the American Academy of Political and Social Science* 140 (November): 52–64.

———. 1935. "Race Prejudice in the Administration of Justice." *American Journal of Sociology* 41, no. 2 (September): 212–17.

———. 1950. "The Uniform Criminal Statistics Act." *Journal of Criminal Law and Criminology* 40, no. 6 (March-April): 679–700.

Sengoopta, Chandak. 2003. *Imprint of the Raj: How Fingerprinting Was Born in Colonial India.* New York: Macmillan.

Shannon, Claude. 1949. "The Mathematical Theory of Communication." In C. Shannon and W. Weaver *The Mathematical Theory of Communication,* by C. Shannon and W. Weaver, 3–91. Urbana: University of Illinois Press. Originally published as "A Mathematical Theory of Communication," 1948.

———. 1956. "The Bandwagon." *IRE Transactions on Information Theory* 2, no. 1 (March): 3.

Sheehey, Bonnie. Forthcoming. "Algorithmic Paranoia: The Temporal Governmentality of Predictive Policing." *Ethics and Information Technology.*

Singer, Peter. 2011. "Visible Man: Ethics in a World without Secrets." *Harper's,* August, 31–36.

Skocpol, Theda. 1992. *Protecting Soldiers and Mothers: The Political Origins of Social Policy in the United States.* Cambridge: Harvard University Press.

Smith, Gavin. 2016. "Surveillance, Data, and Embodiment." *Body & Society* 22, no. 2: 108–39.

Smith, Willard C., and Isadore S. Falk. 1939. "Social Security Needs for Vital Statistics Records." *American Journal of Public Health* 29, no. 5 (May): 452–57.

Stanley, Sharon. 2017. *An Impossible Dream? Racial Integration in the United States.* New York: Oxford University Press.

Stark, P. 1924. "Development of the Real Estate Profession." *National Real Estate Journal* 25, no. 2 (January 28): 35–38.

Stephenson, Gilbert T. 1914. "The Segregation of the White and Negro Races in Cities." *South Atlantic Quarterly* 13 (January): 1–18.

Sterne, Jonathan. 2012. *MP3: The Meaning of a Format.* Durham: Duke University Press.

Stiegler, Bernard. 2010. "Memory." In Mitchell and Hansen 2010, 64–87.

———. 2013. "The Most Precious Good in the Era of Social Technologies." In *The Unlike Us Reader: Social Media Monopolies and their Alternatives,* edited by G. Lovink and M. Rasch, 16–30. Amsterdam: Institute of Network Cultures.

Straus, Nathan. 1952. *Two-Thirds of a Nation: A Housing Program.* New York: Knopf.

Supervision. 1939. "Social Security Gigantic Job." *Supervision* 1, no. 3 (April): 10–12.

Susman, Warren I. 1984. "'Personality' and the Making of Twentieth-Century Culture." In *Culture as History: The Transformation of American Society in the Twentieth Century,* 271–85. New York: Pantheon.

Syeed, Nafeesa, and Elizabeth Dexheimer. 2017. "The White House and Equifax Agree: Social Security Numbers Should Go." *Bloomberg News,* October 3, 2017, https://www.bloomberg.com/news/articles/2017-10-03/white-house-and-equifax-agree-social-security-numbers-should-go.

Symonds, Percival M. 1924. "The Present Status of Character Measurement." *Journal of Educational Psychology* 15, no. 8: 484–98.

———. 1927. "The Present Status of Character Measurement." *Journal of Educational Psychology* 18, no. 2: 73–87.

Szreter, Simon. 2007. "The Right of Registration: Development, Identity Registration, and Social Security—A Historical Perspective." *World Development* 35, no. 1: 67–86.

———. 2012. "Registration of Identities in Early Modern English Parishes and amongst the English Overseas." In Breckenridge and Szreter 2012, 67–92.

Szreter, Simon, and Keith Breckenridge. 2012. "Recognition and Registration: The Infrastructure of Personhood in World History." Editors' introduction to Breckenridge and Szreter 2012, 1–36.

Taeuber, Karl E., and Taeuber, Alma F. 1965. *Negroes in Cities: Residential Segregation and Neighborhood Change*. Chicago: Aldine.

Taylor, W. S. 1928. *Morton Prince and Abnormal Psychology*. New York: D. Appleton.

Tenen, Dennis. 2017. *Plain Text: The Poetics of Computation*. Stanford: Stanford University Press.

Terman, Lewis. 1916. *The Measurement of Intelligence: An Explanation of and a Complete Guide for the Use of the Stanford Revision and Extension of the Binet-Simon Intelligence Scale*. Boston: Houghton-Mifflin.

Terranova, Tiziana. 2004a. "Communication beyond Meaning: On the Cultural Politics of Information." *Social Text* 22, no. 3 (Fall): 51–73.

———. 2004b. *Network Culture: Politics for the Information Age*. London: Pluto Press.

Theerman, Paul. 2010. "Julia Lathrop and the Children's Bureau." *American Journal of Public Health* 100, no. 9. (September): 1589–90.

Thomas, William I., and Florian Znaniecki. 1918–20. *The Polish Peasant in Europe and America: Monograph of an Immigrant Group*. 5 vols. Boston: Gorham.

Thompson, Debra. 2016. *The Schematic State: Race, Transnationalism, and the Politics of the Census*. Cambridge: Cambridge University Press.

Thompson, Slason. 1921. "Reasonable Statistics Would Save Millions to Carriers." *Railway Review* 69 (November 5): 610–11.

Thorndike, E. L. 1918. "The Nature, Purposes, and General Methods of Measurements of Educational Products." In *Seventeenth Year Book of the National Society for the Study of Education, Part II*, edited by G. Whipple. Bloomington: Public School Publishing.

Tobey, James A. 1925. *The Children's Bureau: Its History, Activities, and Organization*. Institute for Government Research, Service Monographs, No. 21. Baltimore: Johns Hopkins University Press.

Toffler, Alvin. 1970. *Future Shock*. New York: Random House.

Torpey, John. 2000. *The Invention of the Passport: Surveillance, Citizenship, and the State*. Cambridge: Cambridge University Press.

Turing, Alan. 1950. "Computing Machinery and Intelligence." *Mind* 59, no. 236 (October): 433–60.

US Census Bureau. 1903a. *Legislative Requirements for Registration of Vital Statistics: The Necessity for Uniform Laws, Methods, and Forms*. Bureau Pamphlet 100. Washington: US Census Office.

———. 1903b. *Practical Registration Methods*. Washington: Census Bureau.

———. 1903c. *Registration of Births and Deaths: Drafts of Laws and Forms of Certificates, & Information for Local Officers*. Bureau Pamphlet 104. Washington: US Census Office.

———. 1906. *Extension of the Registration Area for Births: A Practical Example of Cooperative Census Methods as Applied to the State of Pennsylvania*. Washington: Bureau of the Census.

———. 1908. *Legal Importance of Registration of Births and Deaths: Report of Special Committee on Vital Statistics to the Conference of Commissioners on Uniform State Laws*. Washington: Government Printing Office.

———. 1917. *Birth Statistics for the Registration Area of the United States, First Annual Report, 1915*. Washington: Government Printing Office.

———. 1918. *Birth Statistics for the Registration Area of the United States, Second Annual Report, 1916*. Washington: Government Printing Office.

———. 1930a. *Birth, Stillbirth, and Infant Mortality Statistics for the Birth Registration Area of the United States, Thirteenth Annual Report, 1927*. Washington: Government Printing Office.

———. 1930b. *Birth, Stillbirth, and Infant Mortality Statistics for the Birth Registration Area of the United States, Fourteenth Annual Report, 1928*. Washington: Government Printing Office.

———. 1936. *Birth, Stillbirth, and Infant Mortality Statistics for the Continental United States, the*

Territory of Hawaii, the Virgin Islands, Nineteenth Annual Report, 1933. Washington: Government Printing Office.

US Children's Bureau. 1913. *Birth Registration: An Aid in Protecting the Lives and Rights of Children, Necessity for Extending the Registration Area.* Monograph No. 1. Washington: Government Printing Office.

———. 1916. *Birth-Registration Test: Explanations and Suggestions Addressed to the Chairmen and Members of Committees.* 2nd ed. Washington: Government Printing Office.

———. 1919. *An Outline for a Birth-Registration Test.* Misc. Series No. 12, Bureau Pub. No. 54. Washington: Government Printing Office.

———. 1937. *The Children's Bureau: Yesterday, Today, and Tomorrow.* Washington: Government Printing Office.

US Children's Bureau, Department of Health & Human Services. 2012. *The Children's Bureau Legacy: Ensuring the Right to Childhood.* Washington: Government Printing Office.

US Federal Housing Administration. 1936a. *Second Annual Report of the Federal Housing Administration for the Year Ending December 31, 1935.* Washington: Government Printing Office.

———. 1936b. *Underwriting Manual: Underwriting and Valuation Procedure Under Title II of the National Housing Act.* Washington: Government Printing Office.

———. 1938. *Underwriting Manual: Underwriting and Valuation Procedure Under Title II of the National Housing Act.* Washington: Government Printing Office.

US Federal Housing Administration/Works Progress Administration. 1935. *Technique for a Real Property Survey: Part I: Survey Procedure.* Washington: Government Printing Office.

US Public Health Service (prepared by Asst. Surgeon General Trask of USPHS). 1914. *Vital Statistics: A Discussion of What They Are and Their Uses in Public Health Administration.* Washington: Government Printing Office.

Van Boskirk, R. L. 1939. "Business Ingenuity Does Miracle Job." *Nation's Business* 27, no. 7 (July): 20–22.

Van Ingen, Philip. 1935. "The Story of the American Child Health Association." *Child Health Bulletin* 11, nos. 5 and 6 (September–November): 149–92.

Vismann, Cornelia. 2008. *Files: Law and Media Technology.* Translated by G. Winthrop-Young. Stanford: Stanford University Press. Originally published as *Akten: Medientechnik und Recht,* 2000.

Von Foerster, Heinz, ed. 1952. *Cybernetics.* New York: Josiah Macy Jr. Foundation.

Wark, McKenzie. 2004. *A Hacker Manifesto.* Cambridge: Harvard University Press.

Watters, Ethel M. 1924. "Democracy and the Individual: Maternity and Child Welfare in the United States." *Child Health Magazine* 5, no. 3 (March): 104–9.

Weaver, Warren. 1949a. "The Mathematics of Communication." *Scientific American* 181, no. 1 (July): 11–15.

———. 1949b. "Recent Contributions to the Mathematical Theory of Communication." In *The Mathematical Theory of Communication,* by C. Shannon and W. Weaver, 94–117. Urbana: University of Illinois Press.

Webb, C. 1925. "The Appraisal of City Property." In *Real Estate Handbook,* edited by B. Snyder and W. Lippincott. New York: McGraw-Hill.

Weber, Max. 2004. "Politics as a Vocation." In *The Vocation Lectures,* edited by D. Owen and T. Strong, translated by R. Livingston, 32–94. Indianapolis: Hackett. Originally published in 1919.

Wellbery, David. 1990. Foreword to Kittler 1990, vii–xxxiii.

Whipple, Guy Montrose. 1910. *Manual of Mental and Physical Tests.* Baltimore: Warick & York.

———. 1914–15. *Manual of Mental and Physical Tests.* 2 vols. Baltimore: Warick and York.

Wiener, Norbert. (1948) 1961. *Cybernetics: Or Control and Communication in the Animal and the Machine*. 2nd ed. Cambridge: MIT Press.

———. (1950) 1988. *The Human Use of Human Beings: Cybernetics and Society*. New York: De Capo.

Wilbur, Cressy Livingston. 1906. "How Can Pennsylvania Secure Effective Registration of Mortality Statistics?" Published as appendix B to US Census Bureau 1906, 8–36.

———. 1913. "Hindrances to the Extension of Uniform Methods for Vital Statistics in the United States." *American Journal of Public Health* 3, no. 12 (December): 1253–61.

———. 1916. *The Federal Registration Service of the United States*. Washington: Government Printing Office.

Williams, Kim. 2006. *Mark One or More: Civil Rights in Multiracial America*. Ann Arbor: University of Michigan Press.

Wilson, Stephen. 1998. *The Means of Naming: A Social and Cultural History of Personal Naming in Western Europe*. London: University College London Press.

Winthrop-Young, Geoffrey. 2011. *Kittler and the Media*. Cambridge: Polity.

Witte, Edwin E. 1962. *The Development of the Social Security Act*. Madison: University of Wisconsin Press.

Wood, David Murakami. 2007. "Beyond the Panopticon? Foucault and Surveillance Studies." *Space, Knowledge and Power: Foucault and Geography*, edited by J. Crampton and S. Elden, 245–63. Aldershot: Ashgate.

Woodbury, Robert M. 1922. "The Decline in Infant Mortality: Figures for 1915 to 1920 in the U.S. Birth Registration Area." *Mother and Child* 3, no. 3 (March): 121–24.

Woodworth, Robert Sessions. 1921. *Psychology: A Study of Mental Life*. New York: Henry Holt.

———. 1933. *Adjustment and Mastery: Problems in Psychology*. New York: Century.

———. 1939. "Autobiography." In *Psychological Issues: Selected Papers of Robert S. Woodworth*, 1–25. New York: Columbia University Press. Originally published in 1932.

Woolley, Suzanne. 2017. "Want to Ditch Social Security Numbers? Try Blockchain." *Bloomberg*, October 9, 2017. https://www.bloomberg.com/news/articles/2017-10-09/want-to-ditch-social -security-numbers-try-blockchain.

Wright, Richard. (1945) 1951. *Black Boy: A Record of Childhood and Youth*. New York: Signet Books.

Wyatt, Birchard E., and William H. Wandel. 1937. *The Social Security Act in Operation: A Practical Guide to the Federal and Federal-State Social Security Program*. Washington: Graphic Arts Press.

Yates, JoAnne. 1989. *Control through Communication: The Rise of System in American Management*. Baltimore: Johns Hopkins University Press.

———. 2005. *Structuring the Information Age: Life Insurance and Technology in the Twentieth Century*. Baltimore: Johns Hopkins University Press.

Yerkes, Robert M. 1921. *Psychological Examining in the United States Army*. Vol. 15 of the Memoirs of the National Academy of Sciences. Washington: Government Printing Office.

Young, Iris Marion. 2000. *Inclusion and Democracy*. Oxford: Oxford University Press.

Zack, Naomi. 2003. "Reparations and the Rectification of Race." *Journal of Ethics* 7, no. 1: 139–51.

Zangerle, John. 1924. *Principles of Real Estate Appraising*. Cleveland: Stanley McMichael.

Zielinski, Siegfried. 2006. *Deep Time of the Media: Toward an Archeology of Hearing and Seeing by Technical Means*. Translated by G. Custance. Cambridge: MIT Press. Originally published as *Archäologie der Medien: Zur Tiefenzeit des technischen Hörens und Sehens*, 2002.

Zohlberg, Aristide. 1997. "The Great Wall against China: Responses to the First Immigration Crisis, 1885–1925." In *Migration, Migration History, History: Old Paradigms and New Perspectives*, edited by J. Lucassen and L. Lucassen, 291–315. New York: Peter Lang.

Zuberi, Tukufu. 2001. *Thicker Than Blood: How Racial Statistics Lie*. Minneapolis: University of Minnesota Press.

Index

The irony of a manual index in an age of relentless machine searchability could not be lost on the writer; accordingly the index was constructed less as an exhaustive database of terms (which most readers already have access to via typical text search functionality) and more with an eye to mapping and cataloging key concepts, their key locations, and some of their more obscure references.